CARBOHYDRATE

Edited by **Mahmut Çalişkan, İ. Halil Kavakli**
and **Gül Cevahir Öz**

Carbohydrate
http://dx.doi.org/10.5772/63183
Edited by Mahmut Çalişkan, İ. Halil Kavakli and Gül Cevahir Öz

Contributors

Avital Schurr, Claudia Vega, Marcela Alviña, Hector Araya, Shigesaburo Ogawa, Isao Takahashi, Keith Stine, Lindy Durrant, JiaXin Chua, Cristina Manuela Dragoi, Andreea Arsene, Ion-Bogdan Dumitrescu, Daniela Elena Popa, George Traian Alexandru Burcea Dragomiroiu, Denisa Ioana Udeanu, Alina Crenguta Nicolae, Cristina Elena Dinu-Pirvu, Olivia Timnea, Fabrice Mutelet, El Sayed R.E. Hassan

Published by InTech

Janeza Trdine 9, 51000 Rijeka, Croatia

Publishing Process Manager Edi Lipovic
Technical Editor SPi Global
Cover InTech Design team

First published April, 2017
Printed in Croatia
Legal deposit, Croatia: National and University Library in Zagreb

Additional hard copies can be obtained from orders@intechopen.com

Carbohydrate, Edited by Mahmut Çalişkan, İ. Halil Kavakli and Gül Cevahir Öz
p. cm.
Print ISBN 978-953-51-3069-7
Online ISBN 978-953-51-3070-3

For EU product safety concerns:
IN TECH d.o.o., Prolaz Marije Krucifikse Kozulić 3, 51000 Rijeka, Croatia,
info@intechopen.com or visit our website at intechopen.com.

Meet the editors

Mahmut Çalışkan is a professor of genetics and molecular biology in the Department of Biology at İstanbul University in Turkey. He obtained his first degree, BSc, in Middle East Technical University, Ankara, and attended the University of Leeds, Department of Genetics, England, for his PhD degree. His main research areas include the role of germin gene products during the early plant development and the characterization and biotechnological use of halophilic Archaea.

İ. Halil Kavaklı is a professor of molecular biochemistry in the Department of Chemical and Biological Engineering at Koç University in Turkey. He obtained his BSc degree from Middle East Technical University, Ankara, in the Department of Biology and attended Washington State University, Department of Genetics and Cell Biology, USA, for his PhD degree. He worked as a postdoctoral fellow at the University of North Carolina at Chapel Hill in the Department of the Biochemistry and Biophysics. His main research interests are the carbohydrate metabolism and circadian clock.

Gül Cevahir Öz is a professor in the Department of Biology at İstanbul University. She received her PhD degree in plant physiology from İstanbul University, İstanbul, in 1997. Her present interests include starch synthesis, especially mechanism of ADP-glucose pyrophosphorylase; the role of hormones in the plant development; biochemical and molecular mechanisms of plant tolerance to abiotic stress; and plant biotechnology.

Contents

Preface

Carbohydrates are the most abundant macromolecules on earth and are very important building blocks of life. Among the biomacromolecules, nucleic acids, proteins, and lipids within cells, carbohydrates are considered as molecules that are mainly utilized for energy production, which drives many biochemical reactions within the cells. Over time carbohydrates have served many important cellular functions including cell–cell recognition, bacterial and virus infection, cancer biology, as a protective role (being the primary component of the cell wall) in plant and unicellular organisms, and in regulating the structure of proteins. Therefore, any defect in enzymes that participate in carbohydrate metabolism results in different types of diseases and lack of cell wall, which eventually lead to the death of some organisms. The following are examples of diseases associated with carbohydrate metabolism: glycogen storage diseases, galactosemia, hereditary fructose intolerance, and mucopolysaccharidoses. In addition to their role within the cell, carbohydrates are very important raw material products for various types of industry. Starch, the primary storage form of the glues in plants and some autotrophic unicellular organisms, has many applications and is used as a stabilizer and filler in the textile industry, as a carbon source for the fermenting of microorganisms, and as a sweetener of many food products. Recently, exopolysaccharides, discrete capsules or soluble slimes located outside the cell to protect microorganisms against desiccation, virus attack, and help them attach to surfaces, are being produced for different commercial proposes. Especially, microbial exopolysaccharides are beginning to replace traditional higher plant polysaccharides such as gum arabic, guar gum, and locust bean gum, which are used as thickeners and stabilizers.

The purpose of this book is to provide a glimpse into the various aspects of carbohydrates by presenting the research of a number of scientists who are engaged in the development of new tools and ideas used to reveal carbohydrate metabolism in health and diseases and as a material to mimic the carbohydrate surfaces that take part in molecular recognition, often from very different perspectives. This book covers broad topics including quality carbohydrates for the prevention and therapy of noncommunicable diseases, lactate, glycolysis, biomass in biofuel production, targets for cancer treatment, and biomaterials. We would like to express our deepest gratitude to all authors who contributed to this book by sharing their valuable work with us. We expect that the readership for the book will be broad because of the relationship between the research into carbohydrates and their chemistry. This book should prove useful to students, researchers, and experts in the area of carbohydrates.

Mahmut ÇALIŞKAN and İ. Halil KAVAKLI contributed equally to the book.

Prof. Dr. Mahmut Çalişkan
Istanbul University Faculty of Sciences
Biology Department, Istanbul, Turkey

Prof. Dr. İ. Halil Kavakli
Koc University, College of Science,
Molecular Biology and Genetics, Istanbul, Turkey

Prof. Dr. Gül Cevahir Öz
Istanbul University Faculty of Sciences
Biology Department, Istanbul, Turkey

Chapter 1

Effect of Quality Carbohydrates on the Prevention and Therapy of Noncommunicable Diseases: Obesity and Type 2 Diabetes

Claudia Vega and Marcela AlviñaHector Araya

Additional information is available at the end of the chapter

http://dx.doi.org/10.5772/66702

Abstract

Glycemic index (GI) is defined as "how certain meals raise blood glucose after eating, expressed as a percentage of the area under the glucose response curve when the same amount of carbohydrate was consumed as glucose or bread." Glycemic load (GL) corrects GI according to the quantity of carbohydrates ingested. Both have been related to a higher risk of developing obesity and type 2 diabetes (DM2). High GI meals have been altered to create structurally similar meals with low GI levels. Observational studies and clinical trials have been developed using subjects with DM2 and subjects with obesity undergoing bariatric surgery. It was possible to lower the GI of meals, keeping the sensory properties of the original high GI preparation. Observational studies conducted on DM2 under treatment with metformin have shown associations between GI, GL and glycated hemoglobin. However, the same has not been proven with DM2 individuals under basal insulin therapy. Another observational study in subjects with obesity undergoing bariatric surgery showed that GI affects weight loss after surgery. Regarding experimental studies, a better glucose response has been seen following low GI breakfast intake in DM2 subjects undergoing intensive insulin therapy (IIT).

Keywords: glycemic index, glycemic load, meals, obesity, type 2 diabetes

1. Introduction

The most important macronutrient in the diet is carbohydrates (CHO). Therefore, their quality may significantly affect the health status of population.

A way to evaluate the healthy quality of carbohydrates (CHO) is through its ability to affect the glycemic and insulinemic response. Jenkins et al. proposed the glycemic index (GI), which ranks foods as having a high, moderate, or low glycemic effect, according to their response after food intake [1].

The GI is a methodology in vivo and is obtained by comparing the area under the curve of glycemia produced after the intake of 50 g of available CHO from a standard food to the glycemic curve produced after the intake of the same amount of CHO from an assay food. The resulting value is expressed as a percentage. As a standard, food used is glucose or white bread, corresponding to a value of 100 [2].

The researches have defined values of GI equal to or greater than 70 as high and values equal to or lesser than 55 as low [3, 4]. If it is also considered that high GI diets are associated with putting one at a greater risk for developing obesity and diabetes, then it is concluded that foods over 70 GI are not healthy, while those that have shown a GI lower than 55 are healthy and, therefore, recommended.

The quality and quantity of CHO affect the GI. As per the definition, the GI does not consider the quantity effect, as it only concerns 50 g of CHO from the assay food and the standard food. However, the exact quantity of CHO does not represent the way they are consumed, especially when related to meals with a variety of food.

The glycemic load (GL) quantifies the glycemic effect of the available CHO in a food portion that is generally similar to a normal intake size. The GL is obtained by multiplying the quantity of available CHO from a portion by the GI of the same food and then dividing by 100 [5]. Thus, the GL introduces the quantity variable of CHO and complements the GI.

High GL is expected to produce higher glycemic incursions. Diets characterized by high GL intake in the long term are associated with a higher risk of developing noncommunicable diseases (NCDs), such as type 2 diabetes (DM2) and obesity.

The GI may be affected by a variety of factors, the speed of starch digestion in the small bowel being the most important.

The digestion speed in vitro is a methodology that determines how fast the available starch presented in food is hydrolyzed by the digestive enzymes (alpha-amylase and amyloglucosidase) in an in vitro incubation. Rapidly digestible starch is hydrolyzed after a 20-min period and slowly digestible starch after a period of 120 min. The remnants not digested after this time are referred to as resistant starch [6].

Rapidly digestible starches are gelatinized starches, meaning they are found in food under wet cook and in autoclaved and extruded foods. Slowly digestible starches limit access to hydrolytic enzymes by virtue of their molecular structure. They are found in partially ground grains and seeds and in high-density foods such as pasta and raw meals. Resistant starch is found in raw foods or in whole grains and seeds where starches are physically inaccessible to enzymes. The retrograded starch formed by cooling gelatinized starch is part of this refractory fraction to the enzymatic action. Starch classification according to speed of digestion was an idea proposed by Englyst in his 1992 work *Classification and measurement of nutritionally important starch fractions*, in which he described four types of starch resistance [6].

http://dx.doi.org/10.5772/66702

The culinary techniques are used to prepare foods both in homes and industrially modify the proportions of fast/slow resistance digestive starch [7]. For example, foods that are sliced, chopped, and ground accelerate the speed of enzymatic digestion by allowing for additional surface contact between the enzymes and starch [8]. Cooking, especially in water, allows for the gelatinization of the starch, which significantly decreases the time enzymes take to break down starch and initiate bowel absorption [9, 10]. On the contrary, gelatinized starch, when cooled, produces a retrogradation phenomenon, where a part of starch have become resistant, resulting in a lower impact on the glycemic response [11–13]. Furthermore, the addition of fats and proteins to starch-rich foods decreases the speed of enzymatic digestion [14]. A boiled potato, for instance, has mainly gelatinized starches, meaning that its consumption produces high glycemic curves upon intake, which is then followed by a greater glycemic drop. If oil is added to the same boiled potato, the curve is soft in both its peak and its fall.

The application of diverse culinary techniques is indispensable in that it allows for the consumption of foods that in their raw state cannot otherwise be consumed. On the other hand, this also diversifies the meals in relationship to organoleptic characteristics and increases acceptability. This strengthens adherence to diets, a fundamental issue that allows for the success of, among other things, nutrition therapy in subjects with NCDs.

NCDs are the most important health burden currently in the world. The WHO recently defined obesity as a disease, and it is now considered as of the most important causes of other NCDs such as DM2, hypertension, and cancer, among others. Obesity is the result from a positive energetic balance over time, that is, when more energy is taken than is expended. Energy intake is regulated by many factors, among them dietetic variables such as GI and GL, which take part of an important role in continuing or discontinuing a given intake [15]. The following mechanisms can be used to explain the effect of GI on intake and its potential effects on control of body weight:

- **Release of small bowel peptides related to satiety**: several peptides released from the small bowel (jejunum) in response to the presence of nutrients (especially glucose) signal satiety. Within the identified peptides, GLP-1 has the notable ability to produce satiety signals related to the CHO intake. Slow digestion associated with the intake of slow GI foods will produce a higher contact time between nutrients and bowel cells, as well as an increase in the production of GLP-1 [16].

- **Postprandial hyperinsulinemic effect**: A high GI food may trigger an increment in the postprandial glycemic curve that is two times stronger than that of food containing the same quantity and distribution of macronutrients but with low GI (e.g., white bread versus spaghetti). This hyperglycemic state triggers a hyperinsulinemic effect, which in turn raises the insulin-glucagon rate, resulting in an increase in the anabolic response and the production of expanded glycogenesis and lipogenesis [17].

- **Glucostatic theory**: the beginning of a meal is described as response behavior when the brain detects signs of energy deficiencies, the natural signal corresponds to a decrease of the glycemia. The high glycemic response of high GI meals, followed by an abundant secretion of insulin, would produce a fast fall in the postprandial glycemic curve, which would contribute to the hungry feeling in the short term. While this theory may explain the mechanism of the effect of the GI with regard to the satiety, some research has shown weak associations between the glycemic response and food intake. Therefore, more studies are necessary to support this mechanism [18].

A review of 19 studies encompassing a total of 248 healthy subjects, in which meals with similar amounts of energy and macronutrients, but varying GI levels, were evaluated, proved that at following low GI food intake, satiety and satiation levels increased more than it did following high GI food intake [19]. Another prospective epidemiologic study with a follow-up of 12 years to 89,432 individuals showed that every 10-point GI increase in one's diet led to an average weight gain of 34 g per year [20]. On the other hand, a randomized clinical study that included 34 overweight and obese individuals who consumed either a high or low GI hypocaloric diet over 4 months showed that those receiving low GI diets lost significantly more weight compared to those receiving high GI diets [21].

A review published in 2007 in the *Cochrane* database that looked at six randomized controlled studies totaling 202 obese subjects who were undergoing low GL or low GI diets concluded that low GI diets have a significantly more profound effect on weight loss than do low GL diets [22].

Despite the studies cited above, the relationship between the glycemic response to food intake and weight control is not currently clear due to the lack of studies measuring glycemia, appetite, and food intake in the long term, as well as contradictory findings in researches and reviews [23].

It is known that being overweight is a risk factor for the development of DM2, making this a public health issue. DM2 affects about 366 million people worldwide, a number that is expected to reach 552 million by the year 2030 [24]. As GI and GL play an important role in the metabolic control of subjects with DM2, Ludwig proposed hypothetic metabolic pathways that explain how high GI meals may affect glucose metabolism, irrespective of its effect on the body weight. Experimental studies performed on animals and humans showed that the high postprandial glycemic curve would produce an increase in oxidative stress which directly affected the functionality of beta cells in the pancreas. These studies also showed that the postprandial hyperinsulinemic alone also affects the functioning of the beta cell. Finally, he noted that the rapid fall of the glycemia observed following the quick increase of the glycemic curve as a consequence of high GI food intake would produce the release of contra-regulation hormones in a state similar to fasting, increasing the free fatty acids in plasma which affect the beta cells as well as the insulin receptor. Therefore three conventional mechanisms are described: glucotoxicity, pancreatic overstimulation, and lipotoxicity [17].

A variety of studies indicate that dietary intake of high GI foods is associated with an increase in the risk of developing obesity and DM2 along with cardiovascular diseases seeing as the hyperglycemia and the hyperinsulinemia are related to this kind of food intake [25–27]. On the other hand, Brand Miller et al. developed a meta-analysis to determine the effect of low GI diets on the glycemic control in subjects with DM2, where the subjects studied showed positive effects to fructosamine and glycated hemoglobin (A1c) levels [28].

A systematic review from *Cochrane* that included 11 randomized controlled studies indicated that metabolic control improves significantly as the result of a low GI diet, producing a reduction of 0.5% of A1c, similar to that obtained with drug therapy [29]. This reduction is important, considering that The UK Prospective Diabetes Study (UKPDS) group found that the 1% decrease of A1c is associated with a 37% decrease in the risk of developing microvascular complications [30].

Finally, the postprandial glucotoxicity would stimulate the expression of pro-inflammatory genes through epigenetic mechanisms which respond to particular dietetic intakes, in this case, through the histone methylations which stimulate the inflammatory gene expression related to release of free radicals in the mitochondrial level [31]. The pre-inflammatory state observed in subjects with DM2 would increase insulin resistance and the loss of secretory functions in pancreas beta cells. Additionally, it has been named as a cause related to the development of vascular complications. Therefore, suggested dietetic guidelines would regulate this state throughout nutrition therapy [32].

2. Body: research methods

2.1. Glycemic index and its effect on sensorial acceptability of commonly consumed meals in Chile

There are a variety of factors that affect glycemic index (GI), such as food origin or culinary techniques used in meal preparation. On this basis, it is important to obtain the in vivo GI from commonly consumed meals in order to have more reliable data about quality of carbohydrates (CHO) consumed by the local population.

Chilean typical meals (n = 6) were measured in 10 young and healthy subjects using the techniques described by Jenkins et al. in order to calculate the GI using experimental methods. The results obtained are shown in **Table 1** [33].

The results achieved from meals composed of rice and meat show a significantly lower GI than those found in other studies [34]. It could be explained by the local culinary techniques,

Meals	Glycemic index mean ± SD
Rice-grounded meat	31.4 ± 28.4
Spaghetti-grounded meat	42.0 ± 23.4
Lentil soup	49.3 ± 29.5
Bean-spaghetti	76.8 ± 43.4
Mashed potatoes-grounded meat	51.0 ± 29.2
Carbonada (veggie beef soup)	82.1 ± 48.7

Source: Araya H, Contreras P, Vera G, Alviña M, Pak N. Eur J Clin Nutr 2002, 56:753-9.

Table 1. Glycemic index of commonly consumed meals in Chile.

as rice is fried in oil before boiling, leading to a decrease in the starch's water absorption capacity. On the other hand, the meals made from beans and pasta show much higher value than those already known [34, 35]. The reason that would explain these findings is the practice of overcooking that is commonly implemented in the Chilean kitchen.

On the other hand, a study on the basis of theoretical GI calculated for subjects with type 2 diabetes (DM2) from high acceptability meals commonly consumed in Chile showed that diet GI was 74.9 ± 11.3 in these individuals [36]. According to an unpublished study, the mean acceptability for the same meals was 6.26 on a scale of 1–7, 7 standing for "I like too much." Therefore, planning low GI meals presents a methodological challenge: keeping the organoleptic characteristics similar to those with high GI in order to ensuring good acceptability. With this aim on mind, a current study planned a low GI lunch for each traditional high GI lunch (n = 10) in an effort to keep the meals similar. These homologue meals were cooked with essentially the same ingredients but using different culinary techniques or including different

	High glycemic index lunches			Low glycemic index lunches		
Main food of main course	**Description**	**GI[1]**	**CHO[2] (g)**	**Description**	**GI**	**CHO (g)**
Tubers (potatoes)	Celery with boiled corn kernels *Cazuela de albóndigas* ("veggie and meatball soup": no skin boiled potatoes, short white pasta, and meat) *Macedonia* (fruits: kiwi, banana, and orange)	121.1	52.6	Celery with raw carrot *Cazuela de albóndigas* ("veggie and meatball soup": boiled potatoes with skin, whole rice, and meat) *Macedonia* (fruits: pear, apple, and orange)	54.8	46.3
Cereal (rice)	Boiled vegetables Meat with boiled white rice and vegetables Diet jelly	100.3	51.9	Raw vegetables Meat with fried brown rice and vegetables Diet yogurt	49.0	51.9
Legumes (lentils)	Tomato with boiled corn kernels Lentils with boiled potatoes and grated bread Baked apple	62.1	56.1	Tomato with raw onion Lentil with brown rice Apple with beaten egg white	38.1	54.8
Cereal (wheat)	Cucumber, tomato, and boiled fava beans Spaghetti with Bolognese sauce Boiled pear with diet jam	68.5	59.8	Cucumber and tomato Whole spaghetti with Bolognese sauce Diet yogurt with apple	46.9	50.7
Vegetables (zucchini)	Boiled corn kernels with boiled grated carrots Zucchini pudding with grated white bread Kiwi with orange juice	82.5	47.6	Cabbage with sliced raw carrots Zucchini Soufflé (bread free) Diet yogurt with oat and apple	44.7	44.7

[1]GI: glycemic index.
[2]CHO: carbohydrates.
[3]Unpublished data.

Table 2. High-g3lycemic-index lunches commonly consumed in Chile and their homologue low-glycemic-index lunches[3].

amounts of each ingredient based on the digestive speed of its starch. All preparations consisted of salad, a main course, and a dessert, with all of the components providing between 50–60 g of available CHO together, an amount theoretically calculated using exchange portions and the chemical composition table from Chilean food pyramid [37, 38]. The GI values were obtained from Brand Miller et al. using white bread as standard food [34].

Table 2 (unpublished data) shows some homologue lunches and their GI values. It is possible to identify that the most significant changes to the meals are reflected in alterations made to salads and desserts. The amounts of ingredients in homologue meals were modified at convenience if a high or low GI lunch was being planned. White cereals were replaced for whole-grain cereals; in addition, fruits and tubers with skin were included based on their insoluble fiber content as this type of fiber acts as a physical obstacle to enzymatic action, meaning digestion speed may be slowed down [35]. Whole foods were replaced by sliced, ground, or grated foods in order to achieve lower GI as these culinary techniques increase surface contact between food and digestive enzymes, subsequently increasing speed of starch digestion and thereby increasing the GI of meals [39]. In addition, in an attempt to decrease the speed of digestion, oil was added when the starch of certain foods suffered gelatinization. Furthermore, wet boiled was used to cook high GI meals in a longer period than low GI meals. Conversely, in order to keep the low GI of meals, dry cooking (e.g. baking) or brief boiling was implemented. Additionally, low GI foods replaced certain similar high GI foods; uncooked oats, for example, replaced bread.

The high GI lunches measured between 62 and 121, while the low GI lunches were between 38 and 55.

In order to verify if low GI lunch acceptability equaled that of the high GI lunches, a seven-point hedonic scale was applied among a non-probabilistic sample of 25 women similar to female population with DM2. As noted in **Table 3** (data not published), an average value

Similar lunches[1]	Lunch acceptability		
	High GI[2,3]	Low GI	P[4]
Salad, *Cazuela de Vacuno* (*veggie and meat soap*), fruit with milk dessert	6.5 ± 0.7	6.7 ± 0.6	NS
Salad, *Cazuela de Albóndigas* (*veggie and meat balls soap*), *Macedonia* (variety of Mediterranean fruits)	6.6 ± 0.5	6.5 ± 0.8	NS
Boiled vegetables, fish with potatoes/brown rice, fruit	6.3 ± 0.9	6.2 ± 0.8	NS
Salad, meat with mashed potatoes/mashed potatoes and pumpkins, fruit	6.0 ± 1.0	5.8 ± 0.9	NS
Boiled vegetables/raw vegetables, zucchini pudding, fruit/oat	6.3 ± 0.8	6.6 ± 0.6	NS
Salad, spaghetti, fruit/diet yogurt with fruit	6.2 ± 0.7	6.6 ± 0.7	NS
Salad, chicken with boiled vegetables, fruit/oat	6.5 ± 0.7	5.9 ± 0.8	0.008
Tomato, lentils, apple	6.2 ± 0.6	6.1 ± 1.2	NS
Potatoes salad, fish pudding, milk dessert	6.5 ± 0.6	6.5 ± 0.8	NS
Boiled vegetables/raw vegetables, rice with meat, diet dessert	6.4 ± 0.9	6.5 ± 0.8	NS

[1]Slash (/): divide foods used in low or high GI lunches.
[2]GI: glycemic index.
[3]Data are means ± SD.
[4]Wilcoxon test, significant differences at $p < 0.05$.
[5]Unpublished data.

Table 3. Acceptability of similar high- and low-glycemic-index lunches, according to hedonic scale (n = 20)[5].

of six points was achieved by the lunches evaluated, representing moderate to high acceptability. In most cases, no significant differences were found between high and low GI meals.

2.2. Relationship between glycemic index, glycemic load, and weight control in subjects with obesity

Glycemic index (GI) and glycemic load (GL) are two dietetic factors that affect food intake and, therefore, explain the weight control effect. Subjects undergoing bariatric surgery (BS) generally see a decrease in the rate of weight loss 6 months after surgery and generally start to regain weight one year following surgery [40]. Therefore, it is important to ask the following: Which factors impact the speed of weight loss or weight regain in obese subjects that have undergone BS?

A prospective descriptive study was conducted to evaluate the relationship between the carbohydrate (CHO) intake and weight loss after surgery in individuals undergoing BS. Twenty women undergoing gastric bypass (GBP) (BMI, 42.02 ± 3.9 kg/m^2; age, 36.9 ± 8.4 years) and twenty women undergoing sleeve gastrectomy (BMI, 37.4 ± 2.9 kg/m^2; age: 33.4 ± 8.5 years) were studied 6 and 12 months following surgery. GI and GL averages were assessed using a 3-day dietary record, and the percentages of excess weight loss (PEWL) and weight loss compared to that lost before surgery (ΔW) were measured. GI and GL averages found at 6 months after surgery were 47.6 ± 6.4 and 45.7 ± 16.6 g, respectively. Twelve months following surgery, the GI of the diet was 52.7 ± 6.7 and GL 63.5 ± 22.3 g. Data showed significant weight loss at 6 months after surgery (ΔW, 27.6 ± 4.6%; PEWL, 67.1 ± 14.6%) and at 12 months after surgery (ΔW, 32.3 ± 7.3%; PEWL, 78 ± 20.1%). As suggested in **Figure 1** (unpublished data), the GI diet showed a significant correlation with PEWL at 6 months after surgery. **Figure 2** (unpublished data) describes the association between the GI diet and PEWL at 12 months after surgery.

The GI showed a significant negative correlation with ΔW ($r = -0.42$; $p = 0.008$) six months after surgery, but this correlation is nevertheless annulated 12 months after surgery [41]. No considerable correlation was found when GL and ΔW were analyzed 6 and 12 months after surgery. It would appear that the quality of CHO impact is stronger than CHO quantity over the weight control, according to *Cochrane* review [22].

When performing multiple regression analysis including all independent variables studied (energy sufficiency, CHO, GI, CHO of high GI, GL) as well as dependent variables PEWL and ΔW six and twelve months after surgery, the only variable that showed association was diet GI. This index explains 19% of the variability of PEWL six months after surgery ($r^2 = 0.189$, $p = 0.005$) and 17% of the variability of ΔW ($r^2 = 0.172$, $p = 0.008$). On the other hand, GI explains 11% of the variability of PEWL one year after surgery ($r^2 = 0.106$, $p = 0.04$).

In conclusion, in this group of women undergoing BS that was evaluated 6 and 12 months after surgery (SG and GBP), GI of the diet showed a negative significant association with weight loss, a trend that was most notable 6 months after surgery. On the other hand, the quantity of CHO and GL in the diet did not show any association at all with weight loss in either of the periods studied [41].

http://dx.doi.org/10.5772/66702

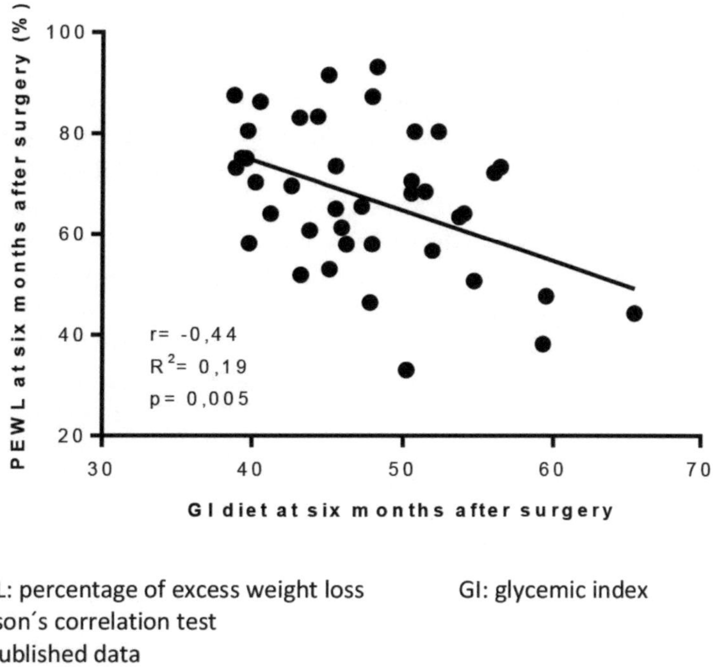

Figure 1. Association between glycemic index diet and percentage of weight loss 6 months after surgery (n = 40)[1].

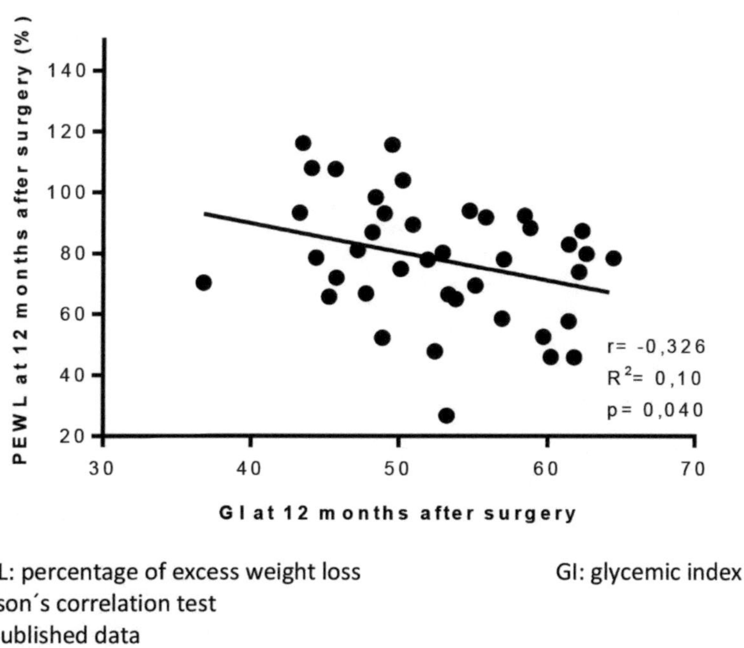

Figure 2. Association between glycemic index diet and percentage of weight loss 12 months after surgery (n = 40)[1].

2.3. Relationship between glycemic index, glycemic load, and metabolic control in subjects with type 2 diabetes

Nutrition therapy plays a key role in metabolic control of type 2 diabetes (DM2) and in the prevention of future diabetes-related complications [42]. Considering the controversial findings shown by previous studies regarding the usefulness of the glycemic index (GI) or glycemic load (GL) in nutrition therapy for DM2 and in order to determine the impact of the quality of carbohydrates (CHO) in the Chilean diet on metabolic parameters in subjects with DM2 under metformin therapy, two transversal descriptive studies were developed. The first one studied 40 individuals with DM2 (age, 58.6 ± 9.5 years; BMI, 32.5 ± 5.8; A1c, 7.08 ± 1.6), in which the subjects were examined through modified frequency food surveys to measure the average amount of usual high GI food portion intake. The findings showed an intake of 403.8 ± 110.7 g of CHO per day. The average GI and GL found were 78.5 ± 3.5 and 317.5 ± 88.3, respectively. The amount of high GI food intake was 21.8 ± 6.5. The association between glycated hemoglobin (A1c) and the amount of high GI food intake is shown in **Figure 3** [43].

The second descriptive study evaluated 108 subjects with DM2 under metformin therapy (age, 53.6 ± 9.45 years; BMI, 30.8 ± 5.8 kg/m^2; A1c, 7.3 ± 1.3 %), all of which were examined through two separate 24-h recall surveys to measure the GI and GL amount in the diet. The findings showed an average CHO intake of 219.8 ± 27.0 grams of CHO per day. Diet GI was 74.9 ± 11.3 and diet GL was 164.0 ± 22.04 g per day. Positive significant associations were found between GI diet and A1c (r, 0.7; $p < 0.05$) and GL and A1c (r, 0.225; $p < 0.05$) [36].

Another research conducted in our department evaluated 40 subjects with DM2 (age, 61.8 ± 8.03 years; BMI, 29.9 ± 4.7 kg/m^2; A1c, 8.6 ± 1.8%) under basal insulin therapy in order

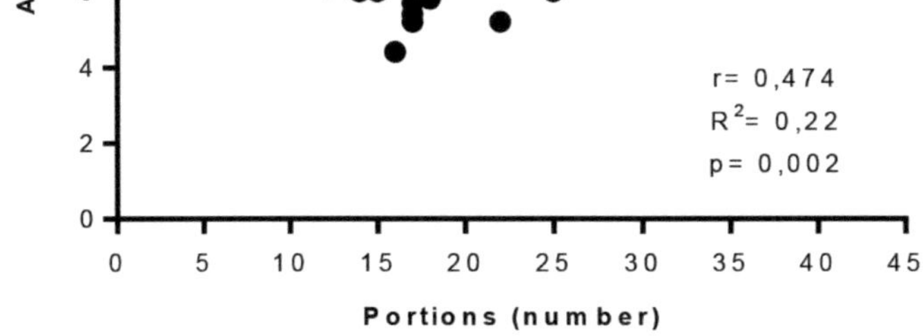

Pearson´s correlation test A1c: glycated hemoglobin
Source: Varela N, Valenzuela K, Vega C. Arch Latinoam Nutr 2012; 62(1):24-29

Figure 3. Correlation between glycemic index food portion intake and A1c in subject with type 2 diabetes under metformin therapy.

to find the relationship between quality of available CHO in every mealtime and metabolic parameters. Individuals were evaluated using 3-day dietary record surveys to identify mealtime GI and GL levels in the diet. Mean fasting glycemia of 163 ± 70 mg dl^{-1} and pre- and postprandial glycemia of 180 ± 73 and 209 ± 83 mg dl^{-1} were found, respectively, and bedtime glycemia was 221 ± 81 mg dl^{-1}. GI and GL of meal time are shown in **Table 4** [44].

The significant associations found in this study are shown in **Figures 4** and **5**. Nonsignificant associations were found for other analyzed variables.

On the basis of these studies, it is possible conclude that, as is widely accepted, the 24-h recall surveys tend to underestimate the food intake and, conversely, the food frequency surveys tend to overestimate the food intake. On the other hand, the GI and GL are associated with metabolic parameters related to DM2 control; nevertheless, this association is weaker when applied in subjects with DM2 under basal insulin therapy.

Variable	Breakfast (n = 40)	Lunch (n = 40)	"Tea time" (n = 38)	Dinner (n = 25)	Snacks (n = 33)
Glycemic index	64.1 ± 8.8^a	44.7 ± 7.6^b	*69.3^a (4.1–4.2)	49.6 ± 9.0^c	45.6 ± 11.8^{bc}
Glycemic load (g)	26.2 ± 11.9^a	17.3 ± 8.1^b	32.6 ± 18.7^c	11.0 ± 8.5^d	13.1 ± 9.3^d

Data are expressed as mean ± SD. *Median (Q1–Q3).
A repeated measure ANOVA was used to achieve differences between meal times.
a, b, c, and d: different upper letters show significantly difference between every mealtime ($p < 0.05$).
Source: Sambra et al. [44].

Table 4. Quality of diet carbohydrates in every meal time.

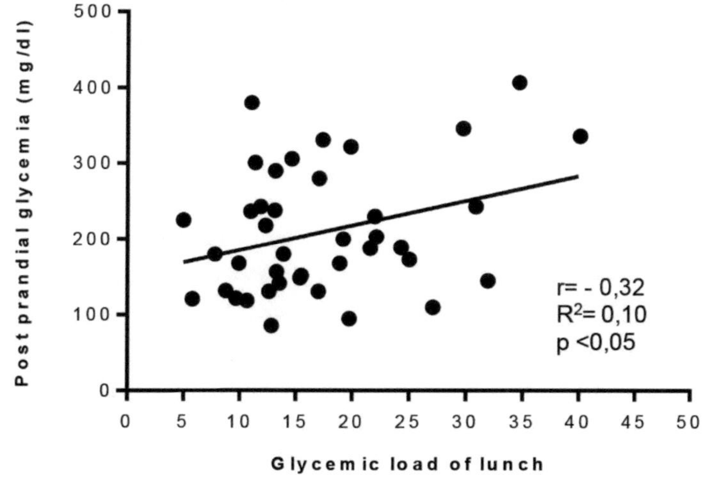

Pearson´s correlation test
Source: Sambra V, Tapia C, Vega C. Nutr Hosp 2015; 31(4):1566-1573

Figure 4. Correlation between postprandial glycemia and glycemic load of lunch in subject with type 2 diabetes under basal insulin therapy (n = 40).

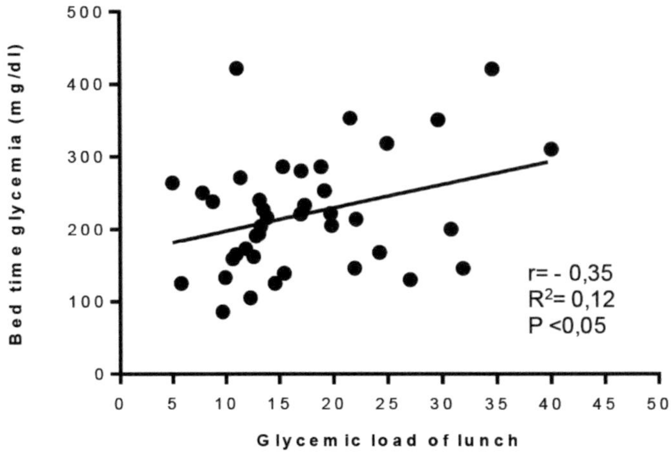

Figure 5. Correlation between bedtime glycemia and lunches glycemic load in subject with type 2 diabetes under basal insulin therapy (n = 40).

2.4. Effect of different GI meal intake over glycemic response and satiety in subjects with obesity, insulin resistance, and type 2 diabetes

2.4.1. Effect of high glycemic index food intake on obese subjects undergoing sleeve gastrectomy

The sleeve gastrectomy is a surgery procedure that aims to decrease gastric capacity [45]. The weight regained one year after surgery is normally found in these kinds of individuals due to an increase in gastric capacity and, hence, a decrease in satiety [40]. Consequently, it is important to consider the qualitative indexes diet, especially those from carbohydrates (CHO) because, according to the last food intake survey (2010–2011) conducted in Chile, this is the main component of energy intake [46]. In order to describe and compare the glycemic response (GR) and glycemic index (GI) in 10 subjects with obesity one year after sleeve gastrectomy versus 10 healthy subjects, a research was conducted. **Figure 6** shows a higher glycemic response 30 min after instant mashed potato intake among subjects undergoing bariatric surgery, as opposed to healthy subjects (SG, 159.8 ± 25.9 mg dl^{-1}; healthy subjects, 135.3 ± 17.3 mg dl^{-1}; $p = 0.023$). However, the GI value calculated from instant mashed potatoes was similar in both studied groups (instant mashed potatoes GI in subjects undergoing SG, 119; instant mashed potatoes GI in healthy subjects, 120; $p = 0.974$) [47].

In conclusion, instant mashed potatoes, despite its similar GI in both groups studied, produce greater glycemic response; therefore, the intake of this food achieves an outcome contrary to the surgery objectives, as it stimulates weight regain.

2.4.2. Effect of white or whole-wheat pasta intake in subjects with insulin resistance

The insulin resistance (IR) is a metabolic condition that allows for the maintenance of normal postprandial glycemia at the expense of a raise in insulin synthesis from the pancreas. In this condition, pro-inflammatory factors, free fatty acids in plasma or ectopic fat, generate damage to the insulin receptor [48]. Additionally, there is a genetic predisposition that contributes

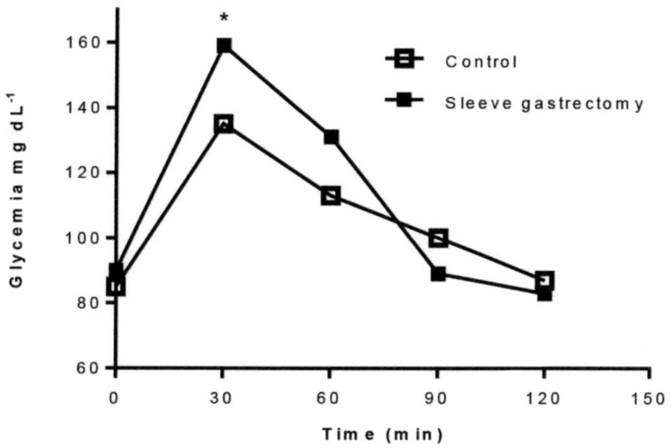

* p<0.05 by one-way ANOVA
Source: Fuentes G, Del Valle M, Vega C. Nutr Hosp 2014; 30(6):1263-1269

Figure 6. Instant mashed potato glycemic response in subject undergoing sleeve gastrectomy (n = 20, control group, 10; Sleeve gastrectomy, 10).

to the development of this condition. In order to assess the glycemic response and GI after white and whole-wheat pasta intake, 10 healthy subjects and 10 individuals with insulin resistance were studied. All participants ate 50 g of available CHO from two types of pasta (white or whole wheat) with equal morphologic characteristics (short pasta) that were subjected to the same culinary technique. No considerable differences in the glycemic response between groups were found in this study for either type of short pasta. On the other hand, **Figure 7** shows a substantial difference found 30 min after intake of white versus whole-wheat pasta which only manifests itself among subjects with insulin resistance (unpublished data) [49].

These findings prove that the insulin-resistance condition meets the metabolic demands of foods, regardless of their GI values. Nevertheless, individuals with insulin resistance would benefit from whole-wheat pasta intake as it invokes a significantly lower glycemic response than does white pasta.

2.4.3. Effect of low and high glycemic breakfast on glycemic response and satiety in subjects with type 2 diabetes under intensive insulin therapy

There is not currently enough consistent evidence about the effect of low GI meal intake in individuals with DM2 under intensive insulin therapy (IIT). A controlled, crossover, single-blind study was conducted in order to determine the effect of low GI breakfast on glycemic response and satiety in 10 subjects (age, 57 ± 8 years; BMI, 34.9 ± 2.0; A1c, 8.9 ± 0.8) with DM2 under IIT. The findings described in **Figure 8** show that low GI breakfast intake achieves a postprandial glycemia significantly lower than high GI breakfast intake (unpublished data) [50].

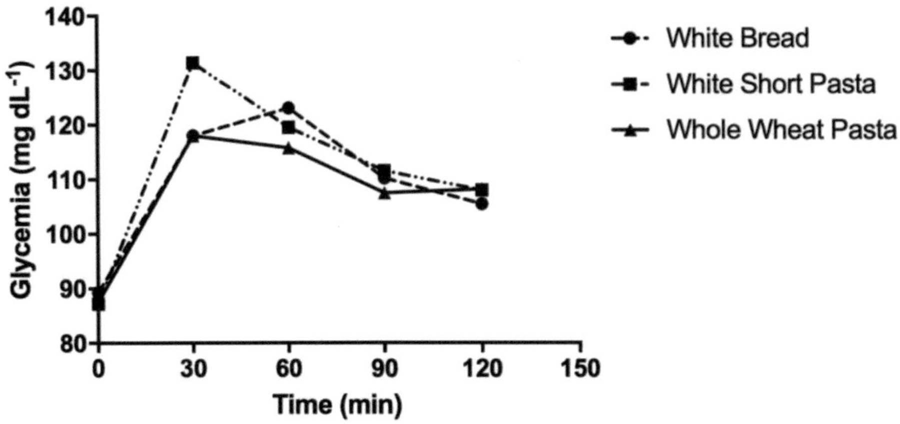

*p<0.05 at 30 minutes after food intake by repeated measures ANOVA.
[1]Unpublished data

Figure 7. Glycemic response after intake of white and whole-wheat short pasta in subject with insulin resistance (n = 10)[2].

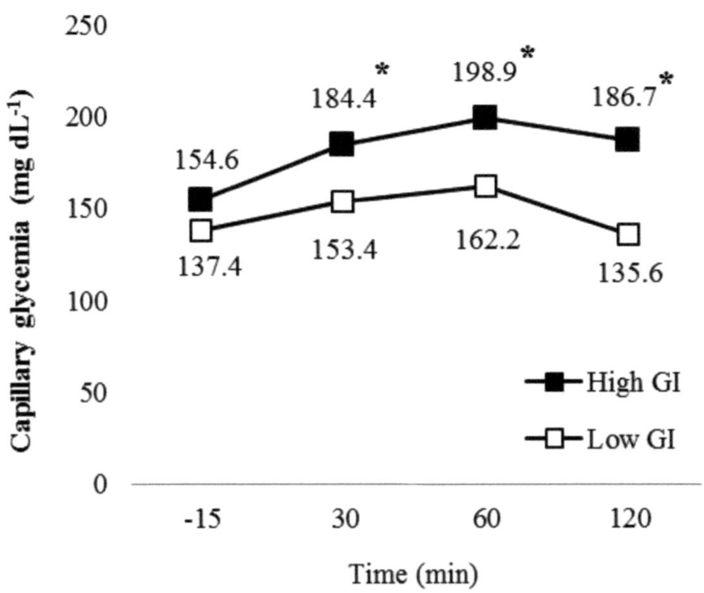

*p<0.05 by paired Student´s test. GI: glycemic index
[1]Unpublished data

Figure 8. Capillary glycemic response after intake of low or high glycemic index breakfast in subjects with type 2 diabetes under intensive insulin therapy (n = 10)[2].

Variable[1,3]	High GI breakfast	Low GI breakfast	P[2]
Hungry	0.0–1.9 (0.80)	0.0–1.0 (0.80)	NS
Satiation	5.2–10.0 (9.20)	9.1–10.0 (9.20)	NS
Fullness	5.1–10.0 (9.00)	8.3–10.0 (9.20)	NS

[1]Data are expressed as range and median. GI, glycemic index.
[2]Paired t-Student's test.
[3]Unpublished data.

Table 5. Satiation in subjects with type 2 diabetes under intensive insulin therapy after low- or high-glycemic-index breakfast intake (n = 10).

Variable[1,3]	High GI breakfast	Low GI breakfast	P[2]
Hungry	0.0–4.0 (1.00)	0.0–5.0 (0.80)	NS
Satiety	2.9–10.0 (8.10)	5.0–10.0 (9.20)	NS
Fullness	4.9–10.0 (8.10)	1.9–10.0 (9.20)	NS

[1]Data are expressed as range and median. GI, glycemic index.
[2]Paired t-Student's test.
[3]Unpublished data.

Table 6. Satiety of subjects with type 2 diabetes under intensive insulin therapy after low- or high-glycemic-index breakfast intake (n = 10).

After evaluating satiety through an analogue visual scale of 10 cm, it was found that subjects who ate a low GI breakfast reported fewer signs of hunger and more fullness immediately after eating than did subjects who consumed a high GI breakfast intake. Nevertheless no significant differences were found (**Table 5**, data no published). Two hours after low GI breakfast intake, individuals stated feeling less hunger and more satiety than high GI intake, although no significant differences were found (**Table 6**, unpublished data).

Regarding the findings described, it is possible to conclude that the low GI breakfast intake achieves significantly lower postprandial glycemic responses, greater satiety, and less hunger than does high GI breakfast intake in individuals with DM2 under intensive insulin therapy.

3. Conclusion: key results

There is currently limited data that describes in vivo GI values of meals commonly consumed in Chile. According to the above findings, it is possible to conclude that local culinary techniques affect GI values, as this study established a difference with the data published until the present. This finding ratifies the importance of reporting local values in order to account authentically data regarding the quality of carbohydrates consumed by population and its corollary effect over health.

With regard to high/low GI homologue meals, it has proven possible to produce a strong impact on the theoretically estimated GI value of various meals by making adjustments regarding the type and proportions of ingredients, as well as culinary techniques applied. These interventions maintained the organoleptic properties of the high GI meal in order to ensure adherence, as was demonstrated by findings acquired.

With regard to the effect of GI and GL on obesity, it is possible to conclude that GI shows a greater impact than does GL in weight loss in obese subjects undergoing bariatric surgery, a trend that is more notable 6 months following surgery than it is 12 months following surgery. In type 2 diabetes (DM2) under metformin therapy, both GI and GL affect the metabolic control in subjects studied. Those findings are different in subjects with DM2 under basal insulin therapy, where GL showed a greater impact on glycemic control that did GI.

Finally, the clinical trials developed establish that high GI meals/food achieves greater glycemic response in subjects with obesity undergoing sleeve gastrectomy, insulin resistance, or DM2 under intensive insulin therapy, especially 30 min following food intake.

Author details

Claudia Vega*, Marcela Alviña and Hector Araya

*Address all correspondence to: claudia.vega@uv.cl

School of Nutrition and Dietetic, Faculty of Pharmacy, Universidad de Valparaíso, Valparaíso, Chile

References

[1] Jenkins D, Wolever T, Taylor R, et al. Glycemic index of food a physiological basis for carbohydrate exchange. Am J Clin Nutr 1981;**34**:362-366.

[2] Mathers J, Wolever T. Digestion and metabolism of carbohydrates. In: Gibney M, Voster H, Kok F, editors. Introduction to Human Nutrition. 1st ed. Oxford, UK. Blackwell Science; 2002. pp. 69-80.

[3] Levitan E, Mittleman M, Hakansson N, Wolk A. Dietary glycemic index, dietary glycemic load, and cardiovascular disease in middle-age and older Swedish men. Am J Clin Nutr. 2007;**85**:1521-1526.

[4] Hare-Bruun H, Flint A, Heitmann BL. Glycemic index and glycemic load in relation to changes in body weight, body fat distribution, and body composition in adult Danes. Am J Clin Nutr. 2006;**84**:871-979.

[5] Brand-Miller JC. Glycemic load and chronic disease. Nutr Rev. 2003;**61**:S49-S55.

[6] Englyst HN, Kingsman SM, Cummings JH. Classification and measurement of nutritionally important starch fractions. Eur J Clin Nutr. 1992;**46**:S33-S50.

[7] Osorio-Díaz P, Bello-Perez LA, Sáyago-Ayerdi SG, Benítez-Reyes MP, Tovar J, Paredes-Lopez O. Effect of processing and storage time on in vitro digestibility and resistant starch content of two bean (*Phaseolus vulgaris* L). J Sci Food Agric. 2003;**83**:1283-1288.

[8] Aguilera JM. Why food microstructure? J Food Eng. 2005;**67**:3-11.

[9] Chung H, Lim HS, Lim S. Effect of partial gelatinization and retrogradation on the enzymatic digestion of waxy rice starch. J Cereal Sci. 2006;**43**:353-359.

[10] Shin SI, Kim HJ, Ha HJ, Lee SH, Moon TW. Effect hydrothermal treatment on formation and structural characteristics of slowly digestible non-pasted granular sweet potato starch. Starch-starke. 2005;**57**:421-430.

[11] Srikaeo K, Furst JE, Ashton JF, Hosken RW. Microstructural changes of starch in cooked wheat grains as affected by cooking temperatures and times. Food Sci Technol-LEB. 2006;**39**:528-533.

[12] Giacco R, Brighenti F, Parillo M, et al. Characteristics of some wheat-based foods of Italian diet in relation to their influence on postprandial glucose metabolism in patients with type 2 diabetes. Brit J Nutr. 2001;**85**:33-40

[13] Rosini MP, Lajolo MF, Menezes WE. Measurement and characterization of dietary starches. J Food Comp Anal. 2002;**15**:367-377.

[14] Berti C, Riso P, Monti LD, Porrini M. In vitro starch digestibility and in vivo glucose response of gluten-free foods and their gluten counterparts. Eur J Nutr. 2004;**43**:198-204.

[15] Galgani J and Ravussin E. Principles of human energy metabolism. In: Ahima R S, editor. Metabolic Basis of Obesity. 1st ed. Philadelphia, USA: Springer; 2011. pp. 1-24.

[16] Gutzwiller JP, Goke B, Drewe J et al. Glucagon-like peptide-1: a potent regulator of food intake in humans. Gut. 1999;**44**(1):81-86.

[17] Ludwig D. The glycemic index physiological mechanisms relating to obesity, diabetes, and cardiovascular disease. JAMA. 2002;**287**(18):2414-2423.

[18] Mayer J. Glucostatic mechanism of the regulation of food intake. N Engl J Med. 1953;**249**:13-16.

[19] Bornet FR, Jardy-Gennetier AE, Jacquet N, Stowell J. Glycaemic response to foods: impact on satiety and long-term weight regulation. Appetite. 2007;**49**(3):535-553.

[20] Du H, van der A DL, van Bakel MME, et al. Dietary glycaemic index, glycaemic load and subsequent changes of weight and waist circumference in European men and women. Int J Obes. 2009;**33**:1280-1288.

[21] Wolever, TMS, Mehling C. Long-term effect of varying the source or amount of dietary carbohydrate on postprandial plasma glucose, insulin, triacylglycerol, and free fatty acid concentrations in subjects with impaired glucose tolerance. Am J Clin Nutr. 2003;**77**(3):612-621.

[22] Thomas D, Elliott EJ, Baur L. Low glycaemic index or low glycaemic load diets for overweight and obesity. In: Cochrane Database of Systematic Reviews. 3rd ed. John Wiley and Sons, Ltd; 2007. p. CD005105.

[23] Anderson GH, Woodend D. Effect of glycemic carbohydrates on short-term satiety and food intake. Nutr Rev. 2003;**61**(Suppl. 5):S17-S26.

[24] Aguiree F, Brown A, Cho NH et al. IDF Diabetes Atlas. 6th ed. Basel, Switzerland: International Diabetes Federation; 2013. 155p.

[25] Salmeron J, Manson JAE, Stampfer MJ. Dietary fiber, glycemic load, and risk of non-insulin-dependent diabetes mellitus in women. JAMA. 1997;**277**(6):472-477.

[26] Maki KC, Rains TM, Kaden VN, Raneri KR, Davidson MH. Effects of a reduced-glycemic-load diet on body weight, body composition, and cardiovascular disease risk markers in overweight and obese adults. Am J Clin Nutr. 2007;**85**(3):724-734.

[27] Liu S, Manson JE, Stampfer MJ, et al. Dietary glycemic load assessed by food-frequency questionnaire in relation to plasma high-density-lipoprotein cholesterol and fasting plasma triacylglycerols in postmenopausal women. Am J Clin Nutr. 2001;**73**(3):560-566.

[28] Brand-Miller J, Hayne S, Petocz P, Colagiuri S. Low-glycemic index diets in the management of diabetes: a meta-analysis of randomized controlled trials. Diabetes Care. 2003;**26**(8):2261-2267.

[29] Thomas D, Elliot E. Low glycaemic index, or low glycaemic load, diets for diabetes mellitus. In: Cochrane Database of Systematic Reviews, editor. The Cochrane Library. John Willey and Sons; 2009.

[30] United Kingdom Prospective Diabetes Study (UKPDS). 13: Relative efficacy of randomly allocated diet, sulphonylurea, insulin or metformin in patients with newly diagnosed non-insulin dependent diabetes followed for three years. Brit Med J. 1995;**310**:83-88.

[31] Gogebakan O, Kohl A, Osterhoff M et al. Effects of weight loss and long-term weight maintenance with diets varying in protein and glycemic index on cardiovascular risk factors. Circulation. 2011;**124**:2829-2838.

[32] Goldberg RB. Cytokine and cytokine-like inflammation markers, endothelial dysfunction, and imbalanced coagulation in development of diabetes and its complications. J Clin Endocrinol Metab. 2009;**94**(9):3171-3182.

[33] Araya H, Contreras P, Vera G, Alviña M, Pak N. A comparison between and in vitro method to determine carbohydrate digestión rate response in Young men. Eur J Clin Nutr. 2002;**56**(8):735-739.

[34] Foster Powell K, Holt S, Brand Miller J. International table of glycemic index and glycemic load values. Am J Clin Nutr. 2002; **76**(1):5-56.

[35] Venn BJ, Mann JI. Cereal grains, legumes and diabetes. Eur J Clin Nutr. 2004;**58**:1443-1461.

[36] Pincheira D, Morgado R, Alviña M, Vega C. Quality of carbohydrates in the diet and its effect on metabolic control in subjects with type 2 diabetes mellitus. Arch Latinoam Nutr. 2014;**64**(4):241-247.

[37] FAO/WHO. Carbohydrate in Human Nutrition: Report of a Joint FAO/WHO Expert Consultation. Paper No. 66 ed. Rome, Italy: FAO. Food and Nutrition; 1998.

[38] Jury G, Urteaga C, Taibo M. Exchange portions and chemical composition of chilean foods pyramid. 1st ed. Santiago, Chile: Institute of Nutrition and Food Thecnologies, Center of Human Nutrition, Faculty of Medicine, Universidad de Chile; 1997. 129p.

[39] Parada JA, Rozowski J. Relationship between glycemic response of starch and its microstructure state. Rev Chil Nutr. 2008;**35**(2):84-92.

[40] Sjöström L, Narbro K, Sjöström CD et al. Effects of bariatric surgery on mortality in Swedish obese subjects. N Engl J Med 2007;**357**:741-752.

[41] Vega C. Effect of carbohydrates and proteins quality on weight loss and metabolic parameters n subjects undergoing bariatric surgery [thesis]. Santiago, Chile: Universidad de Chile; 2013. 68p.

[42] American Diabetes Association. Nutrition therapy recommendations for the management of adults with diabetes. Diabetes Care. 2014;**37**(1):S120-S143.

[43] Varela N, Vega C, Valenzuela K. Relationship of consumption of high glycemic index diet and levels of A1c in subjects with type 2 diabetes under metformin or nutrition therapy. Arch Latinoam Nutr. 2012;**62**(1):23-29.

[44] Sambra V, Tapia C, Vega C. Effect of fragmentation and quality of carbohydrates on metabolic control parameters in subjects with type 2 diabetes under basal insulin therapy. Nutr Hosp. 2015;**31**(4):1566-1573.

[45] Yousseif A, Emmanuel J, Karra E et al. Differential effects of laparoscopic sleeve gastrectomy and laparoscopic gastric bypass on appetite, circulating acyl-ghrelin, peptide YY3-36 and active GLP-1 levels in non-diabetic humans. Obes Surg. 2014;**24**(2):241-252.

[46] Amigo H, Bustos P, Pizarro M. National Survey of Food intake: final report. [Internet]. 1st ed. Santiago, Chile: february 02, 2016. [updated march 02, 2016; cited july 12, 2016]. Available in: http://web.minsal.cl/encadescarga/]

[47] Fuentes G, Deel Valle M, Vega C. Comparative analysis of the glycemic response and glycemic index of instant mashed potatoes in subjects undergoing laparoscopic sleeve gastrectomy and control subjects. Nutr Hosp. 2014;**30**(6):1263-1269.

[48] Matsuzawa Y, Funahashi T, Nakamura T. The concept of metabolic syndrome: contribution of visceral fat accumulation and its molecular mechanisms. J Atheroscler Thromb. 2011;**18**:629-639.

[49] Del Valle M, Teran C. Comparative studies of the glycemic response and glycemic index of two high carbohydrates foods, in subjects with insulin resistance [thesis]. Valparaiso, Chile: Universidad de Valparaíso; 2013. 90p.

[50] Lobos D, Vicuna I. Effect of a sucralose preload and low glycemic index breakfast intake in metabolic post prandial parameters in subjects with type 2 diabetes under intensive insulin therapy [thesis]. Valparaíso, Chile: Universidad de Valparaíso; 2015. 90p.

Chapter 2

Lactate, Not Pyruvate, Is the End Product of Glucose Metabolism via Glycolysis

Avital Schurr

Additional information is available at the end of the chapter

http://dx.doi.org/10.5772/66699

Abstract

Glucose is the monosaccharide utilized by most eukaryotes to generate metabolic energy, and in the majority of eukaryotic systems, glycolysis is the first biochemical pathway where glucose breaks down via a series of enzymatic reactions to produce relatively small amounts of adenosinetriphosphate (ATP). In 1940, the sequence of these glycolytic reactions was elucidated, a breakthrough that was recognized as the very first such elucidation of a biochemical pathway in history. Accordingly, the glycolytic breakdown of glucose ends up either with pyruvate as the final product under aerobic conditions or with lactate, to which pyruvate is being reduced, under anaerobic conditions. Consequently, pyruvate has been designated and is held to be the substrate of the mitochondrial tricarboxylic acid cycle, where it is completely oxidized into CO_2 and H_2O, while lactate has been defined and being held to as a useless dead-end product, poisonous at times, of which cells must discard off quickly. More than four decades after the glycolytic pathway has been elucidated, studies of both muscle and brain tissues have suggested that lactate is not necessarily a useless end product of anaerobic glycolysis and may actually play a role in bioenergetics. These studies have shown that muscle and brain tissues can oxidize and utilize lactate as a mitochondrial energy substrate. These results have been met with great skepticism, but a large number of publications over the past quarter of a century have strengthened the idea that lactate does play an important and, possibly, a crucial role in energy metabolism. These findings have shed light on a major drawback of the originally proposed aerobic version of the glycolytic pathway, that is, its inability to regenerate nicotinamide adenine dinucleotide (oxidized form) (NAD^+), as opposed to anaerobic glycolysis that features the cyclical ability of the glycolytic lactate dehydrogenase (LDH) system to regenerate NAD^+ upon pyruvate reduction to lactate. An examination of scientific investigations on carbohydrate metabolism of brain tissue in the 1920s and 1930s has already revealed that lactate can be readily oxidized. However, due to the prevailing dogma, according to which lactate is a waste product, its oxidation was assumed to be a possible mechanism of elimination. This chapter examines both old and new research data on glucose glycolysis both in muscle and in brain tissues. This chapter consolidates the available data in an attempt to form a more accurate and clear description of this universal and very important bioenergetic chain of reactions.

Keywords: carbohydrate, energy metabolism, glucose, glycolysis, habit of mind, lactate, pyruvate, tricarboxylic acid cycle

1. Introduction

Glucose (D-glucose), also known as dextrose, is a monosaccharide found in its free form in many fruits and also in the blood of humans and other animals. Glucose is combined with fructose to form the disaccharide sucrose (sugar) and is the building block of the most abundant polysaccharides, cellulose, starch and glycogen. In the majority of eukaryotes, from yeasts to humans, glucose is the principal substrate for the production of chemical energy (adenosinetriphosphate ATP), where it is being hydrolyzed via a series of enzymatic reactions, known as glycolysis, to entrap the chemical energy found in glucose chemical bonds.

Glycolysis was the first biochemical metabolic pathway to be elucidated over 75 years ago [1] and thus holds a special place in the annals of our biochemical knowledge. As such, glycolysis has always been described as a pathway that could have two different end products. Under normal aerobic conditions, glycolysis proceeds through nine enzymatic reactions to produce pyruvate; under anaerobic conditions, pyruvate is converted by one additional enzymatic reaction to lactate. The latter has been considered a useless end product, of which tissues must be rid of, as many investigators, then, and even now, held it to be harmful. This description of the glycolytic pathway has stood unchallenged for more than six decades. However, beginning in the 1980s, studies in the fields of both muscle and brain energy metabolism have indicated that lactate is not a useless product of anaerobic glycolysis, but rather a potential important player in energy metabolism in these tissues and possibly others. The present chapter describes the key biochemical and physiological data both from the early days of research on carbohydrate metabolism and those gathered over the past three decades that have challenged the original, dogmatic layout of the glycolytic pathway. Hopefully, this chapter will spur biochemists, physiologists and neuroscientists to consider the reconfiguration of glycolysis as proposed here and elsewhere.

2. Glycolysis circa 1940

In almost every biochemistry textbook published over the past 70 years, glycolysis is described thusly: *"Glycolysis is the sequence of reactions that converts glucose into pyruvate with the concomitant production of a relatively small amount of ATP"* [2]. This usually follows with the qualification that under aerobic conditions, the glycolytic pathway leads up to the tricarboxylic acid cycle (TCA) and the electron transfer chain (ETC), the two biochemical processes responsible for capturing the majority of energy contained in glucose. Thus, under aerobic conditions, pyruvate is the glycolytic product that enters the mitochondria, where through the TCA cycle and the ETC, it is being oxidized to CO_2 and H_2O. In contrast, under anaerobic conditions, such as those existing in working muscles, pyruvate is reduced to lactate.

The elucidation of the glycolytic pathway was completed in 1940, thanks mainly to studies by Meyerhof, Embeden, Parnas, Warburg, Neuberg and Gerty and Carl Cori. It has been the first biochemical pathway to be elucidated, opening the door for future such puzzle solutions and to the field of biochemistry as we know it today. For those who are interested in refreshing their knowledge about the ten or so enzymatic steps of glycolysis and the coenzymes, substrates and products of these steps, any recent biochemistry textbook will do (see also **Figure 1A** and **B**). Nevertheless, despite some uncertainties that have led to unproven assumptions about the role and function of the two alternative glycolytic end products, pyruvate and lactate, the glycolytic pathway has been accepted as originally proposed in 1940. The first nine reactions of glycolysis are summarily listed in **Figure 1A**. These nine reactions end with pyruvate, the product suggested as the substrate for the mitochondrial TCA cycle under aerobic conditions. Since

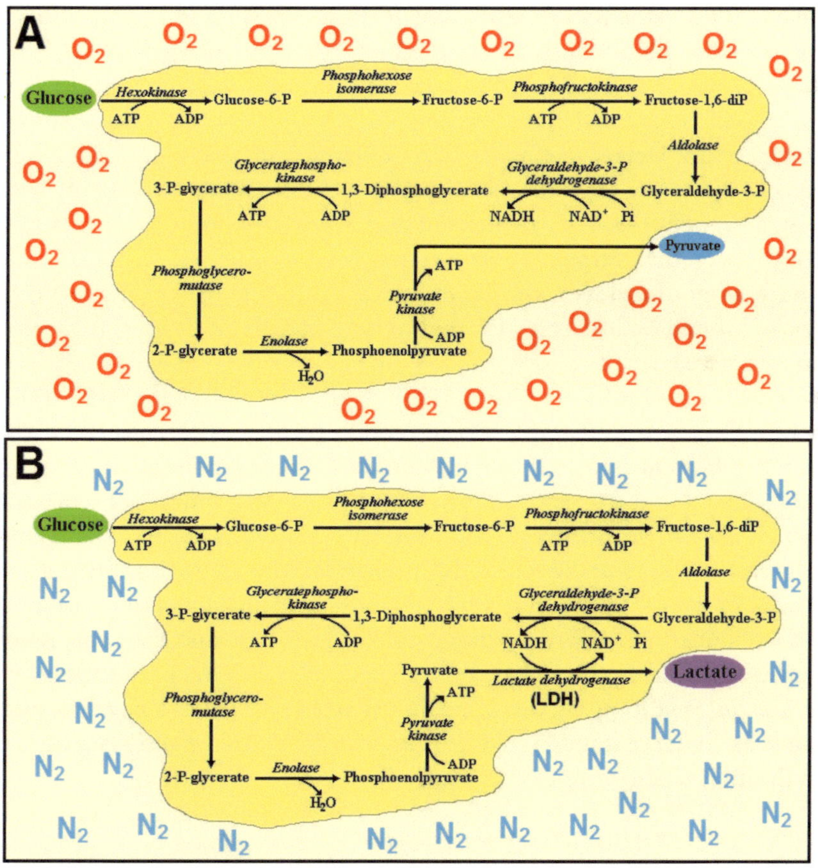

Figure 1. A schematic illustration of the classic glycolytic pathway as originally perceived both under aerobic (A) and anaerobic (B) conditions. Under aerobic conditions, pyruvate is assigned as the end-product of the pathway, while under anaerobic conditions, lactate is the end product. According to this classic concept, NAD^+, an absolutely necessary coenzyme that assures the cyclical nature of glycolysis, cannot be regenerated under aerobic conditions. Only under anaerobic conditions, with the conversion of pyruvate to lactate, NAD^+ is being regenerated. This is one of the main drawbacks of the classical aerobic glycolytic pathway. ATP = adenosine triphosphate; ADP = adenosine diphosphate; NAD^+ = nicotinamide adenine dinucleotide (oxidized form); NADH = nicotinamide adenine dinucleotide (reduced form).

under anaerobic conditions mitochondrial respiration is halted, a 10th reaction was added to the original glycolytic pathway formulation where pyruvate is reduced to lactate by lactate dehydrogenase (LDH, **Figure 1B**). Hence, under anaerobic conditions, glycolysis was postulated to reach a dead-end point.

3. New findings challenge the long-held glycolytic dogma

In 1985, Brooks [3] published results showing that during prolonged exercise of skeletal muscle, lactate is both produced glycolytically and consumed oxidatively. A year later, Fox and Raichle [4] demonstrated "a focal physiological uncoupling between cerebral blood flow and oxidative metabolism upon somatosensory stimulation in humans." Moreover, Fox et al. [5] also showed that "during focal physiologic neural activity, the consumption of glucose is non-oxidative." At the same time, Schurr et al. [6] demonstrated that brain slices in vitro can maintain their normal neuronal function in an oxygen atmosphere with lactate as the sole energy source. Surprisingly, although contracting muscle anaerobic production of lactate has been the dogma ever since Hill's studies in the early 1900s [7–11], when stimulated brain was shown to produce lactate and also utilize it, many scientists exhibited great skepticism [12–17]. Brooks's discovery [3] that skeletal muscle utilizes lactate oxidatively has brought to the fore its own skeptics [18–21]. The finding that activated brain tissue produces lactate [5] should not have been that surprising, since it indicates that activated brain tissue resorts to non-oxidative energy production similar to activated muscle tissue. However, the findings by both Brooks [3] and Schurr et al. [6] that muscle and brain tissues, respectively, utilize lactate as an oxidative energy substrate shook the field of energy metabolism. Consequently, one must wonder why it took over four decades to produce results that challenge the dogma of two separate glycolytic pathways, aerobic and anaerobic. Alternatively, could it be that earlier findings in both muscle and brain tissues had already pointed at the possibility that lactate is more than just a useless end product of glycolysis, but for obscure reasons were ignored? In a review article, Schurr [22] examined the history of carbohydrate energy metabolism from its earlier stages at the end of the nineteenth century to the elucidation of the glycolytic pathway in 1940 and beyond. That review has unearthed some intriguing findings, both about the scientists who were leading the field at the time and the interpretation of their own research data. The scientific debate that ensued following the publications by Brooks [3] and Schurr et al. [6] is still raging on today due, at least in part, to a psychological phenomenon described as "habit of mind" [23] that is known to "afflict" scientists when dealing with a new breakthrough when it appears to contradict common knowledge.

4. The sour reputation of lactate is largely responsible for misconstruing the glycolytic pathway

Sour milk, where lactic acid (lactate) was first discovered, sets the tone for what has become for years to come the negative trademark of this monocarboxylate. Once found in working muscle, lactate was immediately blamed for muscle fatigue and rigor. As early as 1898,

Fletcher [24] demonstrated that lactic acid he used (0.05–5.0%) produced rigor mortis in an excised frog *Gastrocnemius* muscle immersed in it. The higher the lactic acid concentration, the quicker the rigor mortis sets in. Moreover, Fletcher and Hopkins [25] have shown that in the presence of oxygen, the survival of the excised muscle was prolonged and so did the acceleration of the disposal of lactate from it. These researchers highlighted the recognition that the body has the means to rid itself from muscular lactate and that there is ample evidence that such disposal is most efficient under oxidative conditions. Thus, the dogma of lactate as a muscular product responsible for fatigue and rigor, one that aerobic conditions enhance its disposal, was already well entrenched among scientists at the beginning of the twentieth century. It is still entrenched today among athletes and their coaches. Hill [7, 8] went even further than Fletcher by suggesting that the role of oxygen in muscle contracture is twofold, to decrease the duration of heat production and to remove lactate from it. Hill's position and, eventually, the position of the majority of the scientists working in this field of research were that lactate is not a fuel. Hill argued that the measured heat production of lactate oxidation was much lower than the calculated value of its complete combustion. It is somewhat perplexing that a scientist of the stature of Hill would argue that if lactate were a fuel, all the energy of its oxidation would be released as heat. The fact that the measured heat of lactate oxidation was only 12% of the calculated heat production should have indicated to him and others that the majority of the energy released from lactate oxidation, 88% of it, could be a conversion to another form of energy or controlled utilization. The leading investigators in the field at the time actually concluded that lactate is a separate entity from the one that is oxidized during muscle respiration and which yields energy and CO_2. Moreover, they held that the energy yielded in respiration is utilized for lactate disposal.

With such reputation, attempts were made to blame lactate for the racking effects of cocaine use, since increased lactate levels in the blood stream of cocaine users were detected [26] or that increased lactate production is the cause of diabetes' devastating consequences [27]. By the 1920s [28, 29], the central theme of these studies and many others had been muscle tissue and its glycolytic formation of lactate. The process had been postulated to always be anaerobic and mainly through the breakdown of glycogen. In addition, when aerobic oxidation takes place, it occurs only after the muscle contracts and its main purpose is the removal of accumulated lactate and its accompanied acidosis. Furthermore, CO_2 released in the process is due to the acid action on the tissue's bicarbonate. That theme clearly highlights lactate's sour reputation, at least where energy metabolism of muscle is concerned. The relationship between lactate and glycogen in muscle and, eventually, in other tissues, including brain, has been a complicating issue in the understanding of glycolysis. "Otto Meyerhof and Archibald Hill were co-awarded the Nobel Prize in Physiology or Medicine in 1923 for their discovery of the fixed relationship between the consumption of oxygen and the metabolism of lactate in muscle" [22]. While the muscular conversion of glycogen to lactate is still in dispute today [30], both Nobel laureates had a long-lasting influence on this field of research. By the mid 1920s, "blaming" lactate as the culprit for any physiological disorder or abnormal condition had become a "habit of mind" [23]. More details on the tendency of scientists in those days to "demonize" lactate are available [22]. Since the majority of scientists in the field of carbohydrate metabolism in those days studied muscle tissue, their interpretation of and opinions about the results of their studies greatly influenced those who studied carbohydrate metabolism of other tissues, especially brain. Thus, the small scientific community that investigated cerebral glycolysis in the late 1920s and early 1930s adopted the opinions of their peers in the field of muscle glycolysis and accepted the popular dogma, according to which, lactate is a useless end product that the brain eliminates via oxidation. That concept stood against their own notion that the results of their studies could indicate lactate oxidative utilization by brain tissue. While Hill and Meyerhof were the leading scientists in the field of muscle carbohydrate metabolism in the 1920s and 1930s, E.G. Holmes was their counterpart in the field of cerebral carbohydrate metabolism. The latter was joined by his wife, B.E. Holmes, to publish a series of four excellent research papers they titled "Contributions to the study of brain metabolism" [31–34]. First, they showed that brain carbohydrates are not the source of brain lactate; however, the brain is capable of forming lactate from added glucose [31]. In their second study, they determined that brain lactate levels fall when there was a fall in blood sugar level, which results in shortage of glucose in the brain [32]. In the third paper of the series, the Holmes found that brain tissue in room temperature or under anaerobic conditions does not exhibit a significant increase in lactate level or a significant fall in glycogen level, but that under aerobic conditions, lactate rapidly disappears, while glycogen level remains unchanged [33]. Thus, the Holmes established that glucose is the precursor of lactate in the brain and that under aerobic conditions, brain lactate content decreases. Additionally, these investigators showed that brain lactate is formed from glucose supplied by the blood and that its levels rise and fall with blood glucose levels, under both hypo- and hyperglycemic conditions. Moreover, they showed that the diabetic brain is not different from the normal brain, where lactate formation and its removal under aerobic conditions are concerned [34]. By 1929, Ashford joined Holmes and the two were able to demonstrate that the disappearance of lactate and the consumption of oxygen are correlated, which, in essence, indicates an aerobic utilization of lactate by brain tissue. Furthermore, these investigators also showed that sodium fluoride (NaF), the first known glycolytic inhibitor, blocked both glucose conversion to lactate and oxygen consumption. Holmes [35] showed in brain gray matter preparation that oxygen consumption was completely inhibited by NaF in the presence of glucose. However, when lactate was used instead of glucose, oxygen consumption was not inhibited by NaF. Consequently, Holmes concluded that the conversion of glucose to lactate must take place prior to its oxidation by brain gray matter. These results and their straightforward conclusion have been completely ignored for over eight decades. This ignorance is especially glaring when one considers the fact that by the time the glycolytic pathway was elucidated in 1940, Holmes and Ashford papers were already available for at least a decade [35, 36] and should have been taken into account prior to the announcement of that elucidation. Hence, 76 years ago, we could have been presented with somewhat different view of the glycolytic pathway instead of the one in which, depending on the presence or absence of oxygen, ends up with either pyruvate or lactate, respectively. One should be able to confidently postulate such a scenario, since the main players involved in the configuration of the glycolytic pathway were clearly aware of the existence of the TCA cycle [37–40] and its dependence on the end product of glycolysis, one which they assumed to be pyruvate based mainly on Krebs and Johnson's [37] own suggestion that pyruvate is the TCA cycle substrate (see below).

http://dx.doi.org/10.5772/66699

Krebs and Johnson were careful to place a question mark following their suggestion that pyruvate is the TCA cycle substrate. However, the elucidators of the glycolytic pathway took a leap of faith, accepting Krebs and Johnson's suggestion as a fact and an easy choice, when one considers the prevailing dogma of lactate being the anaerobic product of muscle glycolysis and of such bad repute that no one would have considered it to be a substrate for the TCA cycle. Hence, lactate's negative reputation entrenched itself in the minds of the scientists who worked with brain tissue, demonstrated the oxidation of lactate and opined that for glucose to be oxidized, it must be first converted to lactate. Thus, the work by the Holmes couple [31–34], Ashford and Holmes [36] and Holmes and Ashford [41] on brain carbohydrate metabolism has been ignored and remained obscure even today, due mainly to habit of mind [23]. This habit of mind prevents many scientists from accepting more recent data that challenge the old dogma of a glycolytic pathway that has two possible outcomes, aerobic and anaerobic. Nevertheless, we must not forget that in 1940, both the fact that the TCA cycle enzymes are located in mitochondria and the role these organelles play in respiration were unknown. Also unknown at the time was the fact that mitochondria contain in their membrane the enzyme lactate dehydrogenase (LDH), which can convert lactate to pyruvate [42–51]. Ignorance is understandable where the general public is concerned as both coaches and athletes continue, unabated, to blame lactic acid for muscle pain following anaerobic effort, even as recently as during the Rio Olympic games despite the fact that this claim has been refuted [52]. Nevertheless, ignorance cannot explain the persistence of the dogmatic aerobic and anaerobic glycolysis concept among scientists, since the knowledge available today does not support this dogma. Hence, the choice by many scientists to ignore or circumvent this knowledge Is most probably due to habit of mind [23].

5. A single glycolytic pathway with glucose as its substrate and lactate as its end product

The preceding sections have attempted to explain why the pioneers who formulated the glycolytic pathway decided to branch it into two types, aerobic and anaerobic. It is clear from the review of the studies that led to this formulation that these pioneers had to overcome several hurdles while gathering the existing information, including, among others, contradictory results and some unknowns. Nevertheless, their formulation of glycolysis has remained unchanged until this day, regardless of some major predicaments it created as the field of energy metabolism has progressed over the years. Many biochemical pathways have been redrawn as research progresses over time, and yet, the one pathway that has never been subjected to any redrawing throughout its 76 year history has been the glycolytic pathway. The reluctance of many scientists in the field to suggest corrections to or even consider its reformulation is unexplainable. Although many argue that reformulation is unnecessary, the simple fact that "lactate as an oxidative energy substrate" is undisputed should have forced one to reconsider the original, outdated formulation. Most importantly, the originally drawn pathway forces those who object to any reformulation which circumvents the more straightforward one according to which the glycolytic pathway always terminates with lactate production.

Consequently, solutions are being offered for the deficiencies of the old dogma of aerobic glycolysis, that is, its inability to regenerate NAD^+, the coenzyme without which the maintenance of this pathway's cyclical nature is impossible. In contrast, the cyclical requirement of the pathway is met in anaerobic glycolysis upon the conversion of pyruvate to lactate and nicotinamide adenine dinucleotide (reduced form) (NADH) to nicotinamide adenine dinucleotide (oxidized form) (NAD^+) (**Figure 1**). Therefore, aerobic glycolysis, as held today, is not capable of regenerating NAD^+. Although it is unknown how oxygen "converts" anaerobic, lactate-producing glycolysis into an aerobic, pyruvate-producing glycolysis, and no theoretic mechanism has ever been offered for such conversion, it has, somehow, become axiomatic. It stands in complete disagreement with the fact that the glycolytic pathway of erythrocytes, the richest of all tissues in oxygen concentration, produces largely lactate from glucose and only minimal amounts of pyruvate [53]. Despite the fact that red blood cell glycolytic pathway is identical to that of other tissues, it produces lactate, both in the presence and absence of oxygen. However, for an unexplained reason, aerobic glycolysis of all other oxygenated tissues supposedly produces mainly pyruvate. Since erythrocytes lack mitochondria, one should doubt that the addition of mitochondria to erythrocytes in a test tube experiment would somehow change red blood cells' lactate production to pyruvate production. Understandably, this paradox has remained unresolved throughout the second half of the twentieth century. However, the cumulative data gathered since the late 1980s are more than sufficient to suggest that this paradox is actually a misconception. Hence, it is bewildering that the majority of scientists in the field of energy metabolism prefer to accept such a paradox, rather than to correct a deficient formula of this biochemical pathway. Consequently, since the original aerobic glycolysis cannot regenerate NAD^+, investigators had to propose alternative pathways for the production of NAD^+.

The malate-aspartate shuttle (MAS) in brain a major redox shuttle supposedly capable to regenerate NAD^+ when aerobic glycolysis is functional has been proposed as one such alternative [54, 55]. It has been argued that the MAS is a major supplier of NAD^+ in the brain when aerobic glycolysis is operational [56]. Dienel and colleagues have published several studies and reviews over the years adamantly rejecting the postulate that lactate may be utilized oxidatively instead of glucose, since glucose is an obligatory energy substrate in the brain. Dienel [56] argues that lactate aerobic utilization requires a stoichiometric MAS activity to oxidize NADH to NAD^+ by cytoplasmic LDH, ignoring the possibility of lactate oxidation to pyruvate by mitochondrial LDH. Under such circumstances, any NADH is formed in the mitochondria, not in the cytoplasm. LDH localization in the mitochondrial membrane and that mitochondria are capable of utilizing lactate as a substrate of the TCA cycle have been demonstrated by many investigators [42–51, 57]. Hence, the presence of a functional mitochondrial LDH could exclude the need for cytoplasmic MAS to transport NAD^+ into the mitochondria. For those who insist that the original formulation of the glycolytic pathway is correct and accurate, the existence of membranous mitochondrial LDH presents a real dilemma, since one must question the role of such enzyme there, as it is unlikely for the reduction of pyruvate to lactate. Consequently, an aggressive push back was mounted against the findings of Brooks et al. [43], demonstrating LDH presence in mitochondria and its postulated role in lactate oxidation [18–21].

With the abundance of published studies over the past 30 years, all pointing in one way or another at a simpler, straight forward, singular glycolytic pathway, it is of utmost importance

http://dx.doi.org/10.5772/66699

to redefine "glycolysis" as a cytosolic biochemical pathway, of which glucose is its substrate and lactate is always its end product. NAD^+, which is being reduced to NADH during the glycolytic pyruvate formation, is then being regenerated by the glycolytic LDH (cLDH, **Figure 2**) as pyruvate is converted to lactate. That reaction affords this portion of glycolysis its cyclical capacity. Under aerobic conditions, lactate is the main substrate of the TCA cycle and, as such, must be considered as the main molecule coupling between the glycolytic and the TCA cycle pathways, one in the cytosol and the other in the mitochondrion, respectively. Lactate is transported from the cytosol into the mitochondrion via a monocarboxylate transporter (MCT) [58, 59], where it is oxidized to pyruvate by mitochondrial LDH (mLDH, **Figure 2**) and also provides the mitochondrion with NADH. This in turn could circumvent the need for the

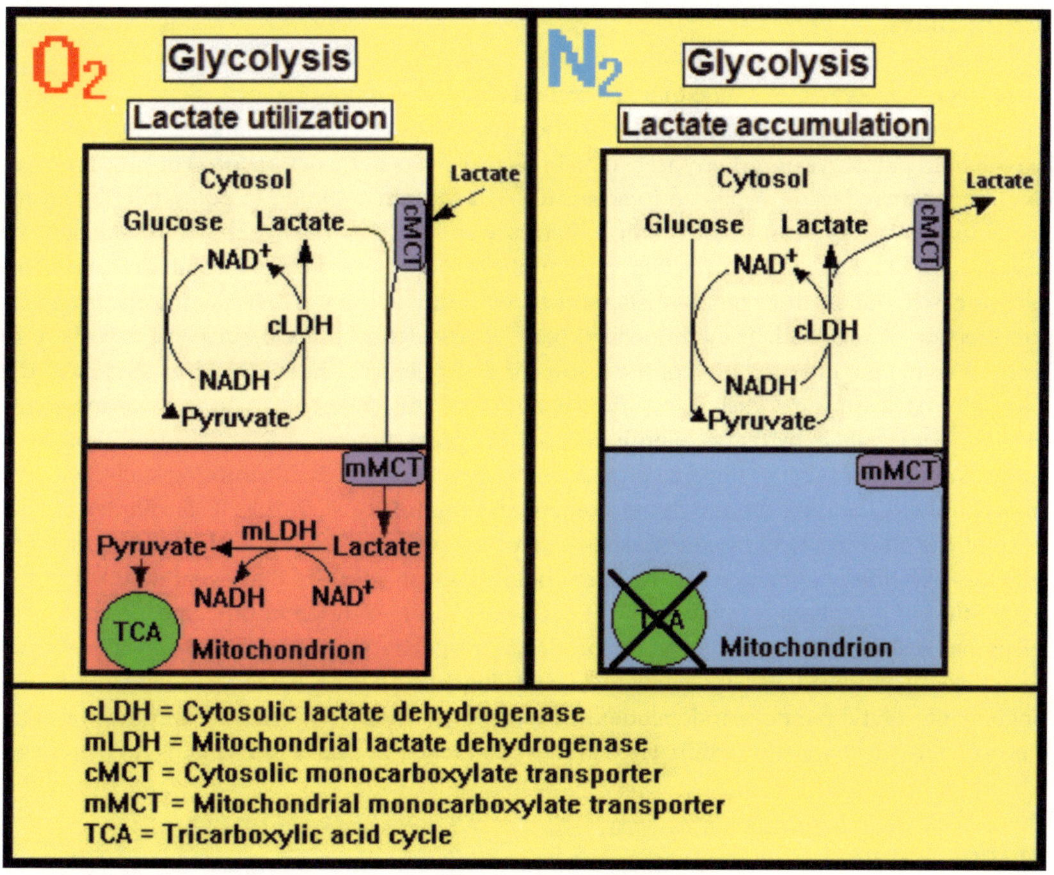

Figure 2. A schematic illustration of the glycolytic pathway as has been proposed based on numerous studies over the past three decades where glycolysis has only one end product, lactate, whether under aerobic or anaerobic conditions. According to this proposed pathway, NAD^+ is being regenerated regardless of the conditions under which glycolysis is operated. Under aerobic conditions (O_2), lactate is being utilized, being the substrate of mitochondrial lactate dehydrogenase (mLDH), which converts it to pyruvate that enters the TCA cycle. Under anaerobic conditions (N_2), lactate is accumulated in the cytosol. NAD^+ = nicotinamide adenine dinucleotide (oxidized form); NADH = nicotinamide adenine dinucleotide (reduced form).

proposed function of the malate-aspartate shuttle (MAS; but see Ref. [60]). Under anaerobic conditions, glycolysis continues to function unabated, resulting in lactate accumulation, as the TCA cycle is nonfunctional (**Figure 2**). When lactate is accumulating, under anaerobic conditions, it becomes upon return to aerobic conditions the principal energy substrate until its levels are falling back to their minimal, normal levels [57, 61–63].

In a recent online Research Topic Ebook published by Frontiers Media SA entitled "Glycolysis at 75: Is it Time to Tweak the First Elucidated Metabolic Pathway in History?" the reader can find research studies, reviews, opinion papers and commentaries highlighting both the growing consensus regarding glycolysis as a pathway with one end product, lactate, and the role of the latter in energy metabolism [22, 60, 64–70].

6. Summary

Lactate is a glycolytic metabolite that has earned a negative reputation ever since its discovery over two centuries ago. Consequently, with the progress of biochemistry and the elucidation of the different pathways of carbohydrate metabolism and bioenergetics, medical or physiological conditions where lactate appeared to accumulate have been assumed to potentially be harmful or damaging. As a result, the medical literature still emphasizes the benefit of reactions or treatments that could minimize lactate concentration. In the early days of carbohydrate metabolism research, the majority of scientists worked with muscle tissue, determining the tone and the direction of this field. They influenced parallel research in other tissues and especially in brain, skewing the interpretation of the results of that research. Therefore, when studies in the mid-1980s have appeared to challenge the prevailing dogma of glycolysis, by postulating a possible role for lactate in oxidative energy metabolism, great number of scientists, then, and even now, allowed their habit of mind to form a barrier that prevents their accepting such a role for lactate, notwithstanding the mounting evidence in support of such role. This chapter details some of the attitudes held by key scientists involved over the years in carbohydrate metabolism research, the possible reasons for them holding those attitudes that eventually led to the description of glycolysis as a biochemical pathway with two different outcomes, aerobic and anaerobic, ending either with pyruvate or lactate, respectively. Also detailed are the original breakthrough studies that have challenged that dogma of glycolysis and instead proposed a singular glycolytic pathway independent of oxygen. Accordingly, this pathway begins with glucose as its substrate and terminates with the production of lactate as its main end product.

Author details

Avital Schurr

Address all correspondence to: avital.schurr@gmail.com

Department of Anesthesiology and Perioperative Medicine, University of Louisville School of Medicine, Louisville, KY, USA

References

[1] Schurr, A. and Gozal, E. (2015) Glycolysis at 75: is it time to tweak the first elucidated metabolic pathway in history? *Front. Neurosci.* 9, 170. doi:10.3389/fnins.2015.00170

[2] Stryer, L. Editor (1995) *Biochemistry*, Fourth Edition, Chap. 19, p. 483. W. H. Freeman and Company, New York.

[3] Brooks, G.A. (1985) Lactate: glycolytic product and oxidative substrate during sustained exercise in mammals—'the lactate shuttle'. In: *Comparative physiology and biochemistry – currecnt topics and trends* Vol A. *Respiration-Metabolism-Circulation*, Ed. R. Gilles, Springer-Verlag, Berlin, pp. 208–218.

[4] Fox P.T. and Raichle, M.E. (1986) Focal physiological uncoupling of cerebral blood flow and oxidative metabolism during somatosensory stimulation in human subjects. *Proc. Natl. Acad. Sci. USA*. 83, 1140–1144.

[5] Fox, P.T. Raichle, M.E. Mintun, M.A. and Dence, C. (1988) Nonoxidative glucose consumption during focal physiologic neural activity. *Science*. 241, 462–464.

[6] Schurr, A. West C.A. and Rigor, B.M. (1988) Lactate-supported synaptic function in the rat hippocampal slice preparation. *Science*. 240, 1326–1328.

[7] Hill, A.V. (1910) The heat produced in contracture and muscular tone. *J. Physiol.* 40, 389–403.

[8] Hill, A.V. (1911) The position occupied by the production of heat, in the chain of processes constituting a muscular contraction. *J. Physiol*. 42, 1–43.

[9] Hill, A.V. (1913) The energy degraded in the recovery processes of stimulated muscles. *J. Physiol*. 46, 28–80.

[10] Hill, A.V. (1914) The oxidative removal of lactic acid. *J. Physiol*. 48, (Suppl) x–xi.

[11] Feldman, I. and Hill, L. (1911) The influence of oxygen inhalation on the lactic acid produced during hard work. *J. Physiol*. 42, 439–443.

[12] Chih, C-P. Lipton, P. and Roberts, E.L. Jr. (2001) Do active cerebral neurons really use lactate than glucose? *Trends Neurosci*. 24, 573–578.

[13] Dienel, G.A. and Hertz, L. (2001) Glucose and lactate metabolism during brain activation. *J. Neurosci. Res*. 66, 824–838.

[14] Chih, C-P. and Roberts, E.L. Jr. (2003) Energy substrates for neurons during neural activity: a critical review of the astrocyte-neuron lactate shuttle hypothesis. *J. Cereb. Blood Flow Metab*. 23, 1263–1281.

[15] Dienel, G.A. and Cruz, N.F. (2004) Nutrition during brain activation: does cell-to-cell lactate shuttling contribute significantly to sweet and sour food for thought? *Neurochem. Int*. 45, 321–351.

[16] Hertz, L. (2004) The astrocyte-neuron lactate shuttle: a challenge of a challenge. *J. Cereb. Blood Flow Metab*. 24, 1241–1248.

[17] Fillenz, M. (2005) The role of lactate in brain metabolism. *Neurochem. Int*. 47, 413–417.

[18] Rasmussen, H.N. van Hall, G. and Rasmussen, U.F. (2002) Lactate dehydrogenase is not a mitochondrial enzyme in human and mouse vastus lateralis muscle. *J. Physiol*. 541, 575–580.

[19] Sahlin, K. Fernstrom, M. Svensson, M. and Tonkonogi, M. (2002) No evidence of an intracellular lactate shuttle in rat skeletal muscle. *J. Physiol*. 541:569–574.

[20] Ponsot, E. Zoll, J. N'Guessan, B. Ribera, F. Lampert, E. Richard, R. Veksler, V. Ventura-Clapier, R. and Mettauer, B. (2005). Mitochondrial tissue specificity of substrates utilization in rat cardiac and skeletal muscles. *J. Cell Physiol*. 203, 479–486.

[21] Yoshida, Y. Holloway, G.P. Ljubicic, V. Hatta, H. Spriet, L.L. Hood, D.A. and Bonen, A. (2007) Negligible direct lactate oxidation in subsarcolemmal and intermyofibrillar mitochondria obtained from red and white rat skeletal muscle. *J. Physiol*. 582, 1317–1335.

[22] Schurr, A. (2014) Cerebral glycolysis: a century of persistent misunderstanding and misconception. *Front. Neurosci*. 8, 360. doi:10.3389/fnins.2014.00360

[23] Margolis, H. (1993) Paradigms and barriers: how habits of mind govern scientific beliefs, The University of Chicago Press, Ltd., London.

[24] Fletcher, W.M. (1898) The survival respiration of muscle. *J. Physiol*. 23, 10–99.

[25] Fletcher, W.M. and Hopkins, F.G. (1907) Lactic acid in amphibian muscle. *J. Physiol*. 35, 247–309.

[26] Underhill, F.P. and Black, C.L. (1912) The influence of cocaine upon metabolism with special reference to the elimination of lactic acid. *J. Biol. Chem*. 11, 235–252.

[27] Ringer, A.I. (1914) Studies in diabetes. I. Theory of diabetes, with consideration of the probable mechanism of antiketogenesis and the cause of acidosis. *J. Biol. Chem*. 17, 107–119.

[28] Hartree, W. and Hill, A.V. (1922) The recovery heat production in muscle. *J. Physiol*. 56, 367–381.

[29] Hartree, W. and Hill, A.V. (1923) The anaerobic processs involved in muscular activity. *J. Physiol*. 58, 127–137.

[30] Shulman, R.G. and Rothman, D.L. (2001) The "glycogen shunt" in exercising muscle: A role for glycogen in muscle energetic and fatigue. *Proc. Natl. Acad. Sci. USA*. 98, 457–461.

[31] Holmes, B.E. and Holmes, E.G. (1925). Contributions to the study of brain metabolism. I. Carbohydrate metabolism. Preliminary paper. *Biochem. J*. 19, 492–499.

[32] Holmes, E.G. and Holmes, B.E. (1925) Contributions to the study of brain metabolism. II. Carbohydrate metabolism. *Biochem. J*. 19, 836–839.

[33] Holmes, E.G. and Holmes, B.E. (1926) Contributions to the study of brain metabolism. III. Carbohydrate metabolism relationship of glycogen and lactic acid. *Biochem. J.* 20, 1196–1203.

[34] Holmes, E.G. and Holmes, B.E. (1927) Contributions to the study of brain metabolism. IV. Carbohydrate metabolism of the brain tissue of depancreatised cats. *Biochem. J.* 21, 412–418.

[35] Holmes, E.G. (1930) Oxidations in central and peripheral nervous tissue. *Biochem. J.* 24, 914–925.

[36] Ashford, C.A. and Holmes E.G. (1929) Contributions to the study of brain metabolism. V. Role of phosphates in lactic acid production. *Biochem. J.* 23, 748–759.

[37] Krebs, H.A. and Johnson, W.A. (1937) The role of citric acid in intermediary metabolism in animal tissue. *Enzymologia*. 4, 148–156.

[38] Krebs, H.A. and Johnson, W.A. (1937) Metabolism of ketonic acids in animal tissues. *Biochem. J.* 31, 64–660.

[39] Krebs, H.A. and Johnson, W.A. (1937) Acetopyruvic acid ($\alpha\gamma$diketovaleric acid) as an intermediate metabolite in animal tissues. *Bichem. J.* 31, 772–779.

[40] Krebs, H.A. Salvin, E. and Johnson, W.A. (1938) The formation of citric and α-ketoglutaric acids in the mammalian body. *Bichem. J.* 32, 113–117.

[41] Holmes, E.G., and Ashford, C.A. (1930). Lactic acid oxidation in brain with reference to the "Meyerfof cycle." *Biochem. J.* 24, 1119–1127.

[42] Brandt, R.B. Laux, J.E. Spainhour, S.E. and Kline, E.S. (1987) Lactate dehydrogenase in rat mitochondria. *Arch. Biochem. Biophys.* 259, 412–422.

[43] Brooks, G.A. Dubouchaud, H. Brown, M. Sicurello, J.P. and Butz, C.E. (1999) Role of mitochondrial lactate dehydrogenase and lactate oxidation in the intracellular lactate shuttle. *Proc. Natl. Acad. Sci. USA*. 96, 1129–1134.

[44] Hashimoto, T. Hussien, R. and Brooks, G.A. (2006) Colocalization of MCT1, CD147, and LDH in mitochondrial inner membrane of L6 muscle cells: evidence of a mitochondrial lactate oxidation complex. *Am. J. Physiol. Endocrinol. Metab.* 290, E1237–E1244.

[45] Schurr, A. and Payne, R.S. (2007) Lactate, not pyruvate, is neuronal aerobic glycolysis end product: an in vitro electrophysiological study. *Neuroscience*. 147, 613–619.

[46] Atlante, A. de Bari, L. Bobba, A. Marra, E. and Passarella, S. (2007) Transport and metabolism of l-lactate occur in mitochondria from cerebellar granule cells and are modified in cells undergoing low potassium dependent apoptosis. *Biochim. Biophys. Acta*. 1767, 1285–1299.

[47] Lemire, J. Mailloux, R.J. and Appanna, V.D. (2008) Mitochondrial lactate dehydrogenase is involved in oxidative-energy metabolism in human astrocytoma cells (CCF-STTG1). *PLoS One*. 3(2): e1550. doi:10.1371/journal.pone.0001550.

[48] Passarella, S. de Bari, L. Valenti, D. Pizzuto, R. Paventi, G. and Altane, A. (2008) Mitochondria and l-lactate metabolism. *FEBS Lett.* 582, 3569–3576.

[49] Gallagher, C.N. Carpenter, K.L.H. Grice, P. Howe, D.J. Mason, A. Timofeev, I. Menon, D.K. Kirpatrick, P.J. Pickard, J.D. Sutherland, G.R. and Hutchinson, P.J. (2009) The human brain utilizes lactate via the tricarboxylic acid cycle: a ^{13}C-labelled microdialysis and high-resolution nuclear magnetic resonance study. *Brain.* 132, 2839–2849.

[50] Elustondo, P.A. White, A.E. Hughes, M.E. Brebner, K. Pavlov, E. and Kane, D.A. (2013) Physical and functional association of lactate dehydrogenase (LDH) with skeletal muscle mitochondria. *J. Biol. Chem.* 288, 25309–25317.

[51] Jacobs, R.A. Meinild, A-K. Nordsborg, N.B. and Lundby, C. (2013) Lactate oxidation in human skeletal muscle mitochondria. *Endocrin. Metabol.* 304, E686–E694.

[52] Pedersen, T.H. Nielsen, O.B. Lamb, G.D. and Stephenson, D.G. (2004) Intracellular acidosis enhances the excitability of working muscle, *Science.* 305, 1144–1147.

[53] Bartlett, G.R. (1959) Human red blood cell glycolytic intermediates. *J. Biol. Chem.* 234, 449–458.

[54] McKenna, M.C. Waagepetersen, H.S. Schousboe, A. Sonnewald, U. (2006) Neuronal and astrocytic shuttle mechanisms for cytosolic-mitochondrial transfer of reducing equivalents: current evidence and pharmacological tools. *Biochem. Pharmacol.* 71, 399–407.

[55] Pardo, B. Contreras, L. Serrano, A. Ramos, M. Kobayashi, K. Iijima, M. Saheki, T. and Satrustegui, J. (2006) Essential role of aralar in the transduction of small Ca2+ signals to neuronal mitochondria. *J. Biol. Chem.* 281, 1039–1047.

[56] Dienel, G.A. (2012) Brain lactate metabolism: the discoveries and the controversies. *J. Cereb. Blood Flow Metab.* 32, 1107–1138.

[57] Schurr, A., and Gozal, E. (2011) Aerobic production and utilization of lactate satisfy increased energy demands upon neuronal activation in hippocampal slices and provide neuroprotection against oxidative stress. *Front. Pharmacol.* 2, 96. doi:10.3389/fphar.2011.00096

[58] Brooks, G.A. Brown, M.A. Butz, C.E. Sicurello, J.P. and Dubouchaud, H. (1999) Cardiac and skeletal muscle mitochondria have a monocarboxylate transporter MCT1. *J. Appl. Physiol.* 87, 1713–1718.

[59] Mowbray, J. (1975) A mitochondrial monocarboxylate transporter in rat liver and heart and its possible function in cell control. *Biochem. J.* 148, 41–47.

[60] Kane, D.A. (2014) Lactate oxidation at the mitochondria: a lactate-malate-aspartate shuttle at work. *Front. Neurosci.* 8, 366. doi:10.3389/fnins.2014.00366.

[61] Schurr, A. Payne, R.S. Miller, J.J. and Rigor, B.M. (1997) Brain lactate, not glucose, fuels the recovery of synaptic function from hypoxia upon reoxygenation: an in vitro study. *Brain Res.* 744, 105–111.

[62] Schurr, A. Payne, R.S. Miller, J.J. and Rigor, B.M. (1997) Brain lactate is an obligatory aerobic energy substrate for functional recovery after hypoxia: Further in vitro validation. *J. Neurochem.* 69, 423–426.

[63] Schurr, A. Miller J.J. Payne R.S. and Rigor, B.M. (1999) An increase in lactate output by brain tissue serves to meet the energy needs of glutamate-activated neurons. *J. Neurosci.* 19, 34–39.

[64] Rogatzki, M.J. Ferguson, B.S. Goodwin, M.L. and Gladden, L.B. (2015) Lactate is always the end product of glycolysis. *Front. Neurosci.* 9, 22. doi:10.3389/fnins.2015.00022

[65] Galow, L.V. Schneider, J. Lewen, A. Ta, T-T. Papageorgiou, I.E. and Kann, O. (2014) Energy substrates that fuel fast neuronal network oscillations. *Front. Neurosci.* 8, 398. doi:10.3389/fnins.2014.00398

[66] Carpenter, K.L.H. Jalloh, I. and Hutchinson, P.J. (2015) Glycolysis and the significance of lactate in traumatic brain injury. *Front. Neurosci.* 9, 112. doi:10.3389/fnins.2015.00112

[67] Brooks, G.A. and Martin, N.A. (2015) Cerebral metabolism following traumatic brain injury: new discoveries with implications for treatment. *Front. Neurosci.* 8, 408. doi:10.3389/fnins.2014.00408

[68] Passarella, S. Paventi, G. and Pizzuto, R. (2014) The mitochondrial L-lactate dehydrogenase affair. *Front. Neurosci.* 8, 407. doi:10.3389/fnins.2014.00407

[69] Chambers, T.W. Daly, T.P. Hockley, A. and Brown, A.M. (2014) Contribution of glycogen in supporting axon conduction in the peripheral and central nervous systems: the role of lactate. *Front. Neurosci.* 8, 378. doi:10.3389/fnins.2014.00378

[70] Goodwin, M.L. Gladden, L.B. Nijsten, M.W.N. and Jones, K.B. (2015) Lactate and cancer: revisiting the Warburg effect in an era of lactate shuttling. *Front. Nutr.* 1, 27. doi:10.3389/fnut.2014.00027

Chapter 3

Monoclonal Antibodies Against Tumour-Associated Carbohydrate Antigens

Jia Xin Chua and Lindy Durrant

Additional information is available at the end of the chapter

http://dx.doi.org/10.5772/66996

Abstract

Glycomic profiling of tumour tissues consistently shows alterations in N- and O-glycosylation profiles of glycoproteins and glycolipids compared to healthy tissues, with important functional implications for cancer cell biology. The overexpression of tumour-associated carbohydrate antigens (TACAs), as a result of aberrant glycosylation in tumours, is usually correlated with poor prognosis and survival of cancer patients. In tumours, TACAs are associated with worse tumour progression than the deletion and inactivation of tumour suppressor genes. The findings of TACAs acting are not merely tumour markers but also constitute part of the machinery in inducing cancer metastasis and invasiveness further strengthen the scientific rationales for immunotherapy targeting TACAs. Despite the attractiveness of the TACAs, there are very few anti-glycan monoclonal antibodies (mAbs), as glycans usually induce low-affinity IgM responses. This chapter provides an overview of TACAs, direct killing anti-glycan mAbs, and introduces two murine mAbs (FG88 mAbs) that recognise Lewis carbohydrate antigens overexpressed on tumour glycoconjugates with high functional affinity. Although the production of anti-glycan mAbs against cancers is not new, the production of high-affinity IgG anti-glycan mAbs is novel. FG88 mAbs definitely have great potential in cancer therapy and serve as valuable tools in glycobiology research.

Keywords: cancer, Lewis carbohydrate antigen, therapeutic monoclonal antibody, oncosis, antibody drug conjugate

1. Introduction

Cell surface glycosylation is a post-translational modification of proteins and lipids, which is universal to all living cells and plays an important role in cell signalling, immune recognition and cell-cell interactions [1]. Monosaccharide units serve as building blocks of glycans

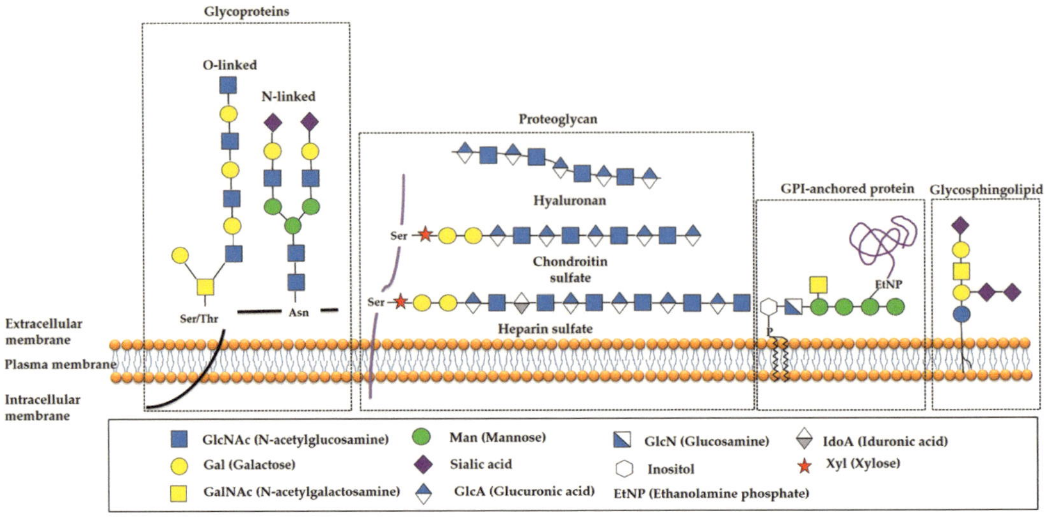

Figure 1. Common glycoconjugates on human cells.

(polysaccharides) that are synthesised by a complex series of post-translational enzymatic steps [1, 2]. There are several main families of glycoconjugates: (1) the Asn-linked (N-linked) and Ser-/Thr-linked (O-linked) oligosaccharides that are present on many glycoproteins, (2) the glycosaminoglycans (GAGs) either as linear-free polysaccharides (such as hyaluronan) or attached to serine residues of proteoglycans (such as heparin sulphate and chondroitin sulphate), (3) the sphingolipids that consist of oligosaccharides linked to ceramide and (4) the glycosylphosphatidylinositol (GPI)-linked proteins, which are proteins that express a glycan chain linked to phosphatidylinositol [2, 3] (**Figure 1**).

2. Glycoproteins

It has been appreciated for some time that protein glycosylation is the most complicated post-translational modification that a protein can undergo [4]. Protein glycosylation is important as it alters the behaviour of proteins, making them more soluble, protecting them from proteolysis, covering antigenic sites and altering the orientation of proteins on cell surfaces [5]. In glycoproteins, the carbohydrate units are linked to the protein backbone by N- and/or O-glycosidic bonds, C-mannosyl bonds, phosphoglycosyl bonds and glypiated linkage (GPI anchor) [4].

N-glycans are covalently attached to the asparagine (Asn) residues of proteins, and the consensus sequence for N-glycosylation is Asn-X-Ser/Thr, where X can be any amino acid except proline. In O-glycosylation, the glycan is attached to the side chains of serine or threonine residues. Unlike N-linked glycosylation, no consensus sequence defining an O-linked glycosylation site has been reported in Ref. [4]. C-mannosylation is a novel type of protein glycosylation, which differs fundamentally from N- and O-glycosylations. It involves covalently

http://dx.doi.org/10.5772/66996

attachment of an α-mannopyranosyl residue to the indole C2 carbon atom of tryptophan (Trp) via a C-C link [6, 7]. The phosphoglycosyl bond is another distinct type of glycopeptide linkage [N-acetylglucosamine (GlcNac), mannose (Man) and fucose (Fuc)] involving an attachment of a carbohydrate to protein via a phosphodiester bond [8]. Another important carbohydrate-protein connection is the GPI anchor. In this connection, mannose is linked to phosphoethanolamine, which in turn is attached to the terminal carboxyl group of the protein [9].

3. Glycolipids

Glycolipids constitute approximately 3% of the outer layer of the plasma membrane, and they are composed of a lipid tail and a carbohydrate head. Glycolipids are classified into three main groups, including glycoglycerolipids, glycosylphosphatidylinositols (GPI) and glycosphingolipids (GSLs), based on the type of lipid component. Of these glycolipids, GSLs are the ones that are most widely overexpressed on tumours [10].

GSLs are ubiquitous membrane constituents, which are embedded in the cell plasma membrane [11]. Ninety percent of mammalian GSL biosynthesis begins with the synthesis of glucosylceramide (GlcCer), which is a key precursor of the glycosphingolipid series [12]. The process takes place on the cytosolic face of the Golgi complex via the action of the Type I transmembrane protein glucosylceramide (GlcCer) synthase [13], which transfers a glucose (Glu) residue to ceramide in the β-glycosidic linkage [14, 15]. Galactose is then added to GlcCer to generate lactosylceramide (LacCer) by β-1,4-galactosyltransferases in the lumen of the Golgi apparatus [10]. Further glycosylation steps are catalysed by different glycosyltransferases

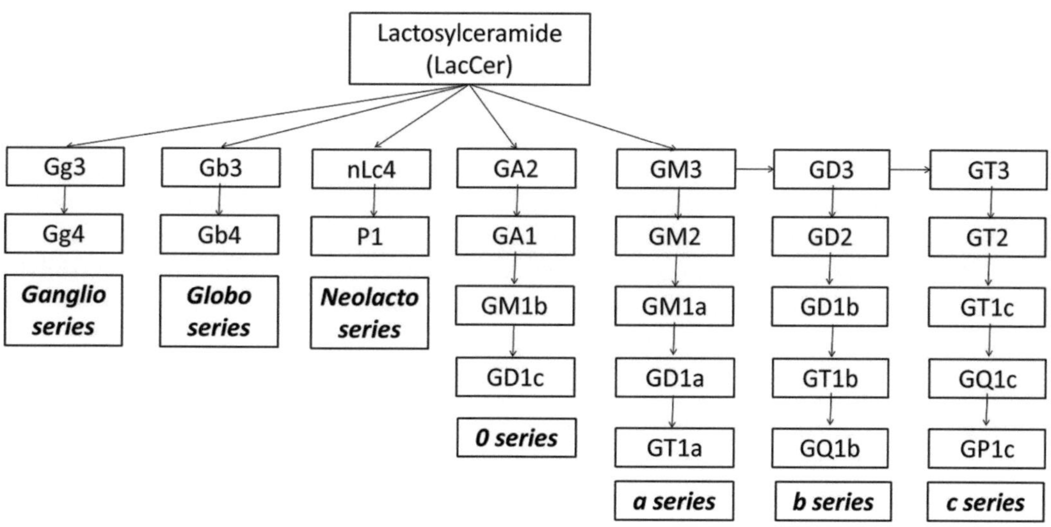

Figure 2. Synthetic pathways for the major GSL species. Lactosylceramide provides the branch point for different GSL series.

with different specificities and result in the generation of more complex GSLs [16]. The major GSL series are defined by their internal core carbohydrate sequence. They are the ganglio-series (galNacβ1-4 gal), globo-series (galα1-4 gal), lacto-series (galβ1-3glcNacβ1-3 gal) and neolacto-series (galβ1-4glcNacβ1-3 gal). LacCer provides the branch point for the synthesis of all these GSL series (**Figure 2**) [12].

4. Aberrant glycosylation in cancers

Aberrant glycosylation has been described as one of the hallmarks of cancer. Aberrant glycosylation of proteins and lipids during malignant transformation leads to the overexpression of tumour-associated carbohydrate antigens (TACAs) [17]. Evidence is accumulating that TACAs have contributed to various aspects of cancer development and progression, including proliferation, invasion, angiogenesis and metastasis [2, 18]. Thus, studying the mechanisms and consequences of variations in glycosylation associated with cancers will provide crucial insight into cancer progression. Importantly, TACAs are overexpressed mostly on tumour cell surface, making them potential diagnostic markers and ideal therapeutic targets [19].

In general, aberrant glycosylation in cancers is due to the following changes: (1) under- or overexpression of glycosyltransferases, (2) altered expression of glycoconjugate acceptor [20], (3) altered sugar nucleotide transporter activity [21] and (4) improper function of the Golgi structure [22].

4.1. Altered glycosylation patterns in cancer

In cancers, the expression of glycosyltransferases is often deregulated. For example, N-acetylglucosaminyltransferase V (GnT-V), which catalyses the formation of β-1,6-GlcNAc branching structures, is expressed only at very low level in normal mammary gland. However, in cancer, the expression of GnT-V has been upregulated, resulting in highly branched N-glycan structures which were found to be associated with cancer growth and metastasis [20, 23–25]. Overexpression of beta-1,3-N-acetylglucosaminyltransferase 8 (β3GnT8) results in increased levels of polylactosamine structures in colorectal carcinoma [26], and upregulation of N-acetylglucosaminyltransferase III (GnT-III) increase bisected N-glycans in liver cancer [27, 28]. Aberrant expression of alpha-N-acetylgalactosaminide alpha-2,6-sialyltransferase 1 (ST6GalNAC-I) in breast cancer resulted in the sialylation of Tn antigen to form the sialyl-Tn (STn) antigen [29]. Altered expression of fucosyltransferases is responsible for the aberrant expression of Lewis carbohydrate antigens such as Lewis a (Le^a), sialyl-Lewis a (SLe^a), Lewis x (Le^x) and Sialyl-Lewis x (SLe^x) in many types of tumours [30–32].

Incomplete glycosylation is another abnormal feature found in human cancer. The expression of truncated O-glycans such as Tn, STn and T antigens had been reported in a wide range of tumours [33–36]. Incomplete glycosylation of these truncated glycans is due to the defects in the secretary pathway organelles (endoplasmic reticulum and Golgi) [33], the absence of glycosyltransferases responsible for the generation of core glycan for chain elongation [37] and the overexpression of sialyltransferases responsible for the addition of terminal sialic acid (i.e. conversion of Tn to STn antigen) [33, 38, 39]. For example, in neoplastic cells, alterations in glycosylation of O-linked glycans had been shown to affect oligomerisation of cell surface receptors, thus influencing the stimulation of these receptors. Wagner et al. demonstrated the inhibition of complex O-glycan formation (N-acetyl-galactosamine galactose core I structure and its subsequent sialylation), resulted in the impairment of death receptors 4 and 5 (DR4 and DR5), which significantly impact the apoptotic pathway signalling by TNF-related apoptosis-inducing ligand (TRAIL) [40].

In addition, specific changes in O-glycan (GalNAc-Ser/Thr) and N-glycan core structures have been reported to lead to the generation of different core glycans with different degrees of glycan branching [27, 28], which in turn significantly impact the overall glycan structure and function.

4.2. Tumour-associated carbohydrate antigens (TACAs)

4.2.1. Altered sialic acid expression

As early as the 1960s, there was evidence that tumour cells of various origins increased the expression of sialic acids on membrane glycoproteins and glycolipids as well as their secretion into the tumour microenvironment [41–43]. Sialic acids on normal cells are involved in multiple different physiologic processes [44]. However, hypersialylation of tumour cells specifically benefits tumour cell growth, promotes metastases [45, 46] and correlates with a poor prognosis of cancer patients [47].

Sialic acids are nine-carbon backbone α-ketoacidic sugars [3]. In general, sialic acids terminate the outer end of glycans (sialoglycans) via more than 20 distinct Golgi-resident sialyltransferases (ST). This enzymatic process is carried out via their second carbon (C2) to either galactose (α2-3Gal or α2-6Gal), N-acetylgalactosamine (α2-6GalNac) or another sialic acid (α2-8Sia) [48]. Altogether, the different linkages to underlying sugars result in a tremendous diversity of sialoglycans [44]. Sialoglycans are known to participate in cell-cell and cell-extracellular matrix interaction, including adhesion, migration and immune recognition [44]. Sialic acid-binding immunoglobulin-like lectins (Siglecs) are receptor families that specifically recognise sialoglycans. Siglecs can be found on most immune cells, and they can transmit immunosuppressive signals upon binding to sialic acid ligands. Thus, increased expression of siglec ligands by tumour cells could contribute to tumour immune invasion [49].

There are a number of causes of the increase in cell surface sialic acid [50]. Changes to the core structures of N-glycans are one of the most common aberrant glycosylations in cancer. Increased activity of GNT-V (also known as MGAT5) was found to result in larger and more branched N-glycans, thus providing additional acceptors for terminal sialylation [20, 50]. Similarly, carcinomas that overproduce mucins (heavily glycosylated high-molecular-weight glycoproteins, e.g. MUCI and MUC4) which contain aberrant O-linked glycosylation can lead to increased sialylation [51, 52]. Together with increased expression of sialyltransferases [53], these enzymes increase cell surface sialylation and metastatic potential [20]. In addition, tumour cells often overexpress α2-6 sialic acid, mainly due to upregulation of the ST6Gal-I [54–56] or ST6GalNAc sialyltransferases [29, 57] that respectively conjugate terminal sialic acid to N-glycans or O-glycans and glycolipids [53]. Mass spectrometry analysis of human serum sialo-glycoproteins revealed increased expression level of α2-6 sialylation in breast cancer [58] and lung cancer samples [59], whereas α2-3 sialylation was increased in prostate cancer samples [60], malignant brain tumours [61] and ovarian serous carcinomas [62].

Aberrant expression of sialic acids confers major advantages to tumour cells. Therefore, by targeting these, sialoglycans overexpressed on tumours may be highly beneficial.

4.2.2. Altered Lewis carbohydrate antigen expression

Lewis carbohydrate antigens can be found on various glycoconjugates in most human epithelial tissues [63]. They are formed by the sequential addition of fucose onto oligosaccharide precursor chains on glycoproteins or glycolipids through the action of a set of glycosyltransferases [64]. Le^x was reported to be overexpressed in breast and gastrointestinal carcinomas. Normal expression of Le^x is restricted on certain normal epithelial cells including the oesophagus, stomach, small bowel, ciliated epithelium of trachea, bronchus [65, 66] and normal human polymorphonuclear neutrophils (PMNs) [67]. Le^y was reported to be overexpressed on ovarian, breast, prostate, colon and lung carcinomas. Although Le^y expression can be found on both normal and neoplastic tissues, Le^y distribution differs between the two tissue types. Expression of Le^y on normal epithelial tissues is restricted to the secretory borders of epithelial surfaces, making it less accessible to circulating antibodies. Conversely, Le^y expression on epithelial cancer cells occurs on all surfaces including luminal surfaces [65].

Sialylated Lewis carbohydrate antigens such as SLe^a and SLe^x are significantly enhanced in cancer [68, 69]. The expression of these cancer-associated antigens results mainly from the upregulation of sialyltransferases [68]. SLe^a is normally present on the inner surface of the ductal epithelium of a variety of epithelial tissues, which makes it largely inaccessible to antibodies and immune effector cells [70]. SLe^x can be found on granulocytes, normal oral mucosa and breast tissue [71]. Both SLe^a and SLe^x are found to be aberrantly expressed on the surface of a broad range of carcinomas such as breast, ovarian, melanoma, colon, liver, lung and prostate [72]. Overexpression of SLe^x and SLe^a appears to directly correlate with increased metastatic disease and poorer overall survival in patients with colorectal cancer invasion [73]. Similar results were obtained from analysis the combination of SLe^a, SLe^x and Le^y antigens in non-small-cell lung cancer (NSCLC) patients [74].

4.2.3. Altered ganglioside expression

Gangliosides are acidic glycosphingolipids with the presence of at least one sialic acid linked to their oligosaccharide chain [75]. Biosynthesis of gangliosides involves sequential addition of sialic acids to lactosylceramide (LacCer) by ST3Gal V (GM3 synthase), ST8SiaI I (GD3 synthase) and ST8Sia V (GT3 synthase) that leads to the generation of the a-, b-, and c-series ganglioside precursors, respectively, representing the mono-, di-, and tri-sialylated gangliosides (**Figure 3**) [68]. In general, gangliosides are involved in cell-cell recognition or regulation of downstream signalling of various proteins (e.g. insulin, epidermal growth factor and vascular endothelial growth factor receptors) [76, 77].

http://dx.doi.org/10.5772/66996

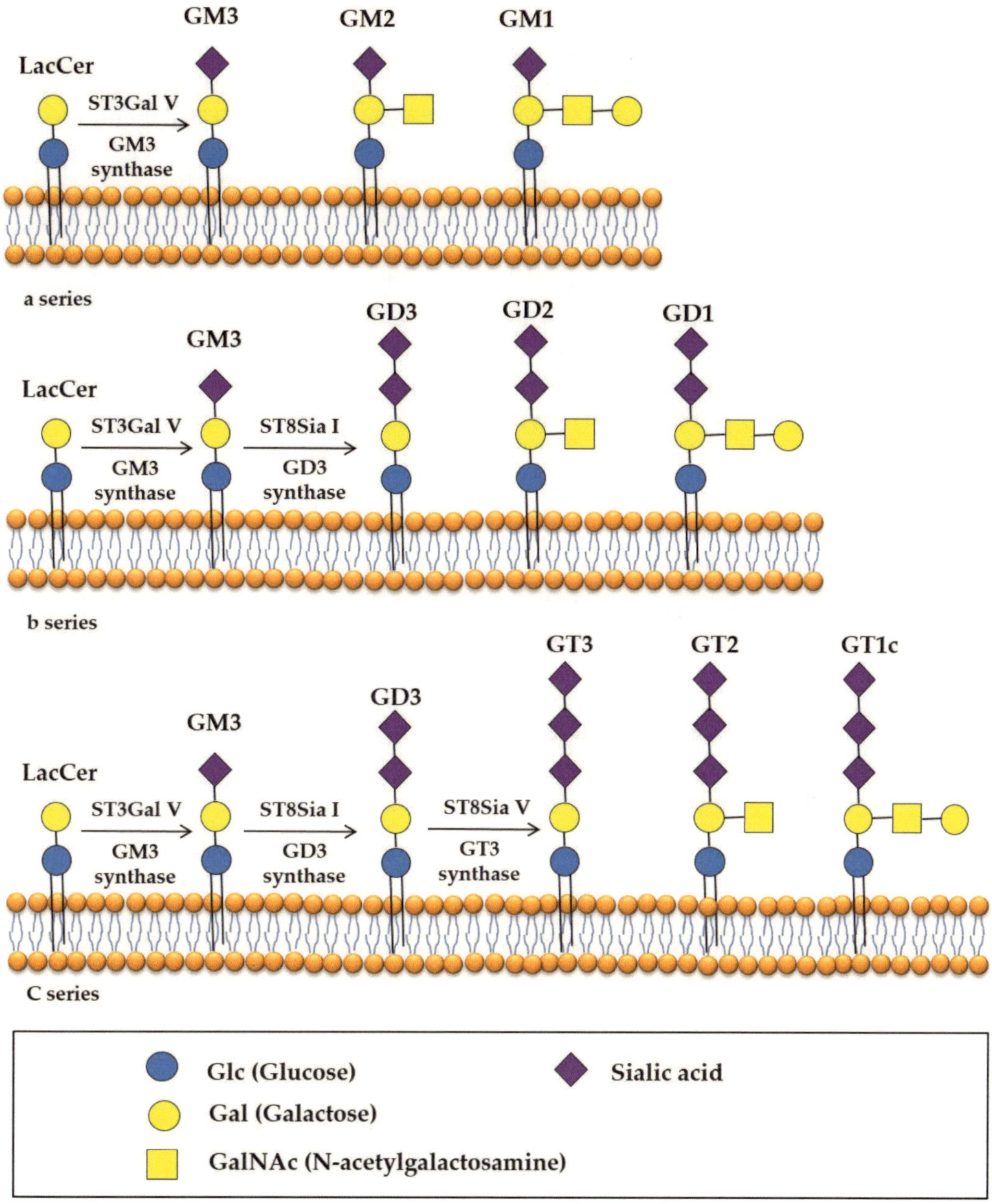

Figure 3. Biosynthesis of gangliosides.

Tumour-associated gangliosides have been suggested as a result of initial oncogenic transformation and play a key role in the induction of invasion and metastasis [75, 78]. Examples of gangliosides that are overexpressed in cancers are GD2 in neuroblastoma [79] and small cell lung cancer (SCLC) [80], GD3 in melanoma [81], GM2 [82] and fucosyl-GM1 in SCLC [81].

Some gangliosides overexpressed in cancers have been identified as adhesion molecules, which promote tumour cell metastasis. Gangliosides such as GD2, GD3 and GT1b form complexes with integrins [83] in melanoma where the terminal sialic acid residues of these gangliosides inhibit cell attachment by abrogating the interaction between integrin $\alpha 5\beta 1$ and fibronectin [84]. Gangliosides on tumour cells also promote metastasis by forming aggregates with peripheral blood mononuclear cells (PBMCs) or platelets. This is due to the interaction between sialic acid moieties on gangliosides with sialic acid-binding proteins named Siglecs (sialic acid/immunoglobulin/lectin) [85], which are expressed on various types of blood cells [86]. These tumour aggregates may induce the blood cells to release factors that activate endothelial cells to elicit cell adhesion molecules (ICAMs, VCAMs, E-selectin and P-selectin), which in turn initiate tumour cell adhesion or invasion [87, 88]. Siglec-7 has been reported to bind preferentially to sialyl-2→6 GalNAc [86]. Interestingly, GD3, GD2 and GT1b share the same sialyl2→6 GalNAc moiety [86], and they are overexpressed on a variety of tumours [89]. Thus, disialo epitopes may promote metastasis by binding to Siglecs expressed on blood cells.

In addition to the cell adhesion function, gangliosides have been reported to act as immune checkpoint molecules to aid in the escape of tumour cells from immune surveillance. Gangliosides released from the active secretion of tumour cells into serum can be taken up by T cells, with ensuing inhibition of T-cell proliferation and activation. The inhibitory action includes the defects in antigen presentation and reduction of cytokine [IFN-gamma (IFN-γ), interleukin-2 (IL-2) and IL-4] production [90–95]. The molecular mechanism of ganglioside-induced T-cell dysfunction was suggested to involve the inhibition of NF-kappa B (NF-κB) activity of T cells via degradation of RelA/p50 dimer and p50/p50 homodimer proteins [96, 97].

4.3. Serum cancer biomarkers

Better survival rates among cancer patients are correlated with earlier detection. The utilisation of serum cancer biomarkers has played a major role in not only early detection of cancer but also prediction of cancer recurrence following initial therapy [98]. However, current clinically approved serum cancer biomarkers are characterised by low sensitivity in detecting cancers [99]. Thus, the development of highly sensitive novel serum cancer biomarkers with better diagnostic and prognostic performance may enhance early detection rates and identification of new targets for anticancer therapy.

4.3.1. α-Fetoprotein (AFP)

α-Fetoprotein (AFP) is a 70 kDa glycoprotein [99], normally only secreted by foetal liver and present in foetal serum [100]. However, under certain pathological conditions, when it is present in adult serum, AFP is associated with cancer. Thus, AFP has been used as a serological marker for the early diagnosis of hepatocellular carcinoma (HCC) [101] and non-seminomatous germ cell tumours (NSGCT) [102].

AFP has a single N-linked oligosaccharide with a biantennary complex-type structure, with altered core fucosylation and terminal sialylation in HCC and NSGCT (**Figure 4**) [102]. Kobayashi et al. reported higher α-1,6-fucosyltransferase (FUT8) expression in HCC tissues than non-cancerous tissues and increased in fucosylation were correlated with HCC

http://dx.doi.org/10.5772/66996

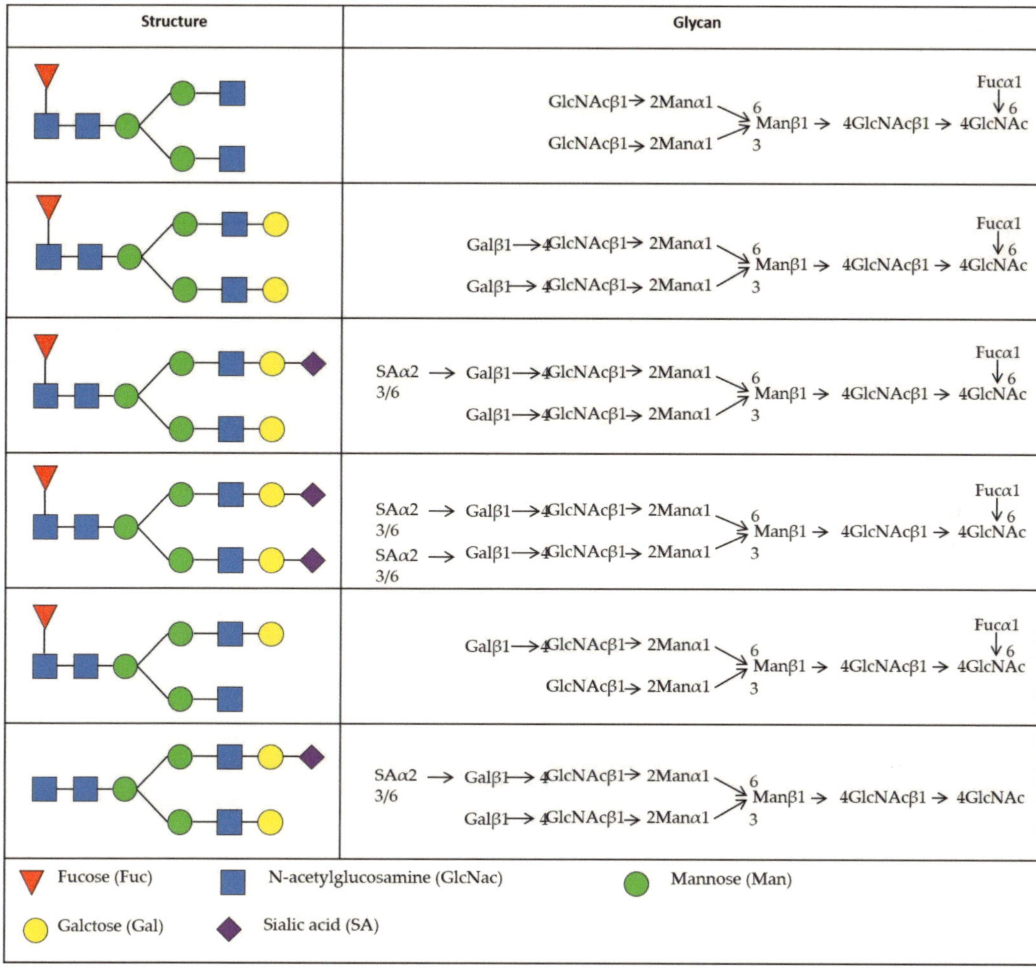

Figure 4. The structures of the N-linked glycans expressed on AFP associated with HCC and NSGCT patients.

progression [101]. In addition to HCC, overexpression of FUT8 in thyroid carcinoma tissue has been linked directly to tumour size and lymph node metastasis [103]. In a study, Osumi and coworkers demonstrated that upregulated of FUT8 expression in cancer cell regulated the expression of E-cadherin. E-cadherin was responsible in the enhancement of cell-cell adhesion, which in turns contributes to the metastatic potential of cancer cells [104].

4.3.2. Prostate-specific antigen (PSA)

Prostate-specific antigen (PSA) is a 28.4 kDa glycoprotein with an N-linked glycosylation site. PSA has been used widely to screen for prostate cancer in men [99]. PSA is normally secreted by the prostate epithelium and periurethral glands. In prostate cancer, disruption of the prostate epithelium leads to the release of PSA into serum [99]. When compared to PSA isolated from healthy individuals, PSA isolated from the serum of prostate cancer patients shown

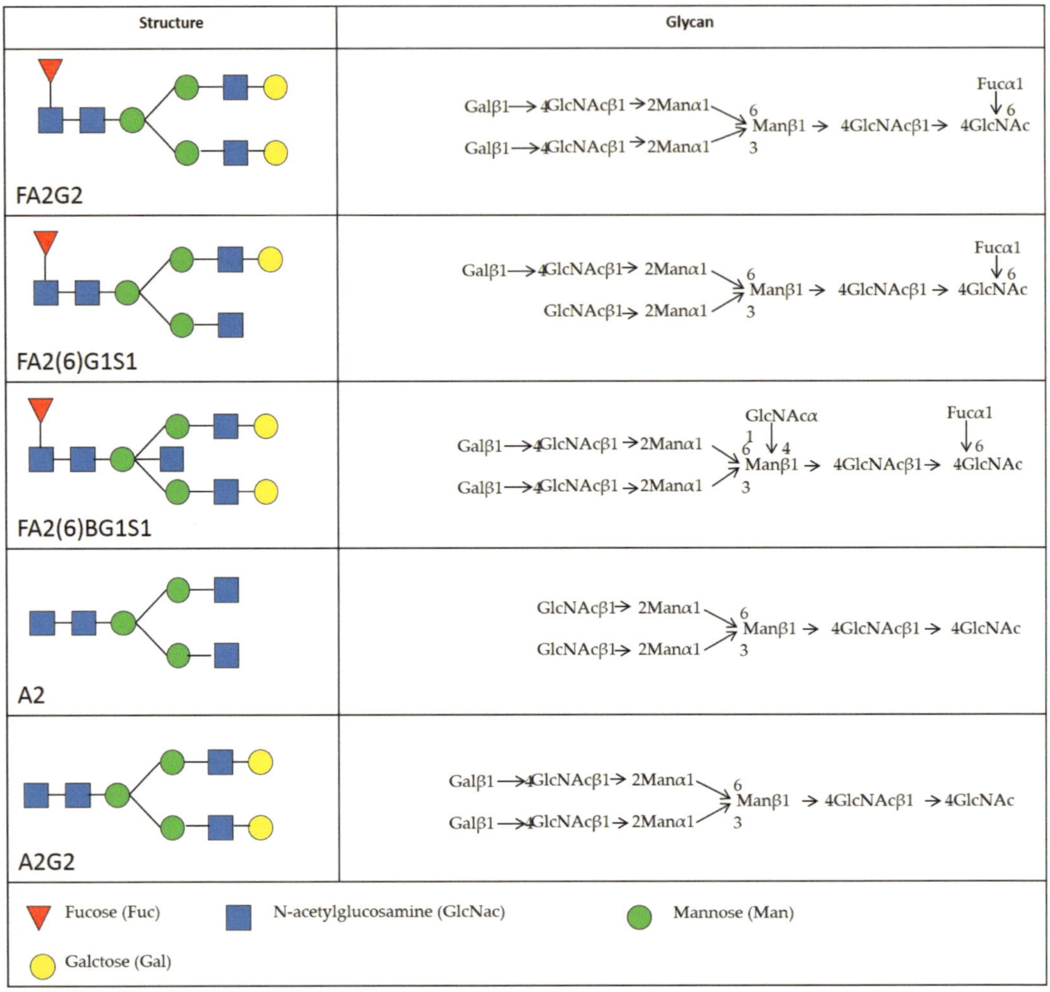

Figure 5. N-linked oligosaccharide structures of PSA elevated in prostate cancer patient serum.

significant higher levels of core-fucosylated biantennary glycans and α-(2,3)-linked sialic acids [105–109]. PSA is composed of several glycoforms [110, 111]. PSA with core-fucosylated biantennary glycans (FA2G2, FA2(6)G1S1 and FA2(6)BG1S1) and terminal α-(2,3)-linked sialic acids (A2 and A2G2) (**Figure 5**) were found elevated in prostate cancer patient serum [60].

4.3.3. Cancer antigen 19-9 (CA19-9)

Cancer antigen 19-9 (CA19-9) corresponds to a carbohydrate structure, sialyl-Lewis a (SLe^a) [99], which is overexpressed on cancer cell surface as a glycolipid and/or as an O-linked glycoprotein [112]. CA19-9 was first characterised by 1116-NS19-9 mAb [113] and has been found primarily in pancreatic and biliary tract cancers [113]. It has been used as a serum biomarker for pancreatic cancer [114]. In neoplastic tissues, epigenetic silencing of the gene for

α-(2,6)-sialyl transferase leads to the abnormal synthesis and accumulation of SLe^a, instead of its normal counterpart disialyl Lewis a (di-SLe^a). SLe^a has been reported to play a crucial role in cancer invasion/metastasis by acting as ligand for endothelial cell E-selectin, which is responsible for cell adhesion [115–118].

It is worth noting that majority of these glycobiomarkers were discovered by generating tumour-specific monoclonal antibodies (mAbs). In addition to aid in the discovery of additional carbohydrate-based biomarkers, these tumour-specific mAbs have great potential in treating neoplastic disease.

4.4. Antibody-based immunotherapy of cancer

Specific recognition and elimination of malignant cells by antibodies were proposed over a century ago [119]. The development of antibody-based therapies for cancer has been the focus of considerable interest for decades. Several criteria have been described for the selection of antitumour mAbs: (1) the mAb binds to cell surface tumour antigen, (2) the mAb binds to tumour antigen at high affinity, (3) the mAb recognises tumour antigen that is overexpressed on tumours but has limited expression on normal tissues, (4) the mAb has potent immune-mediated and non-immune-mediated cytotoxicity effects, (5) the mAb directly kills tumour cells and/or (6) the mAb internalises into target cells so it can delivery toxic payloads. To date, many therapeutic monoclonal antibodies (mAbs) have been developed, mostly against protein antigens, and have proven useful in cancer therapy.

4.5. Anti-glycan monoclonal antibody against cancer

Anti-glycan mAbs have also found usage in clinical applications. Dinutuximab is a chimeric mAb directed against GD2 on neuroblastoma and induced cell lysis via Antibody dependent cellular cytotoxicity (ADCC) and Complement dependent cytotoxicity (CDC) (http://www.fda.gov/). It was approved by FDA in 2015, for use in high-risk neuroblastoma paediatric patients, in combination with granulocyte-macrophage colony-stimulating factor (GM-CSF), interleukin-2 (IL-2) and 13-cis-retinoic acid (RA). GD2 is a disialoganglioside overexpressed on neuroblastoma and melanoma [79, 120] with limited expression on normal neurons in the cerebellum, skin melanocytes and peripheral nerves [121], making it well suited as target for cancer therapy. The FDA approval was based on the findings from a phase III trial, focused on 226 children with high-risk neuroblastoma who had responded to initial treatment [122]. Patients were enrolled to receive either Dinutuximab associated with IL-12, GM-CSF and RA or RA alone, after having responded to the first-line treatment. Three years after treatment assignment, patients received Dinutuximab combination showed superior rates of event-free survival and overall survival when compared to standard therapy.

BR96 is an anti-Le^y mAb. It was shown to induce direct tumour cell death to Le^y-positive tumour cell lines in addition to ADCC or CDC [123]. In a phase I trial, BR96-doxorubicin immunoconjugates showed limited clinical antitumour activity with patients experiencing a clinically significant hypersensitivity reaction to the mouse man [124]. BR96 was conjugated to doxorubicin and docetaxel (also known as SGN-15), and 62 advanced non-small-cell lung cancer patients were treated in a randomised Phase II trial. An increased in survival was reported for patients receiving SGN-15 when compared with patients receiving doxorubicin and docetaxel alone [125]. hu3S193 is a humanised anti-Le^y mab. It contains only 3–5% of murine residues in the antibody variable domain, conferring a low risk of hypersensitivity responses. In a MCF-7 xenograft preventive model, hu3S193 managed to significantly slow tumour growth compared with placebo and isotype-matched control IgG1 antibody [126]. In a phase I trial in 15 cancer patients (six breast, eight colorectal and one non-small-cell lung cancers), hu3S193 showed only minimal toxicity. The biodistribution of indium 111 radio-labeled hu3S193 ((111)In-hu3S193) showed no evidence of normal tissue uptake, but (111) In-hu3S193 uptake was seen in cutaneous, lymph node and hepatic metastases [127].

KM231 is a murine mAb recognising SLe^a. KM231 was observed to react with many human gastrointestinal cancer tissues and could detect shed antigen in the sera of cancer patients. Shitara et al. made KM231-ricin A chain immunotoxin to evaluate the tumoricidal effect of KM231 on ascites and subcutaneous xenograft tumours growing in nude mice. KM231 significantly inhibited the growth of established subcutaneous tumours. This result suggested that it was an effective tumoricidal drug when it was conjugated to cytotoxic reagents [128]. 5B1 (IgG1) and 7E3 (IgM) are other anti-SLe^a mAbs generated by immunising mice with SLe^a-KLH vaccine. Both mAbs are very potent in inducing CDC. Moreover, 5B1 is also highly active in inducing ADCC [129].

NCCT-ST-421 (IgG3) is a murine mAb raised by immunising mice with human gastric cancer xenograft (ST-4). The mAb recognises dimeric Le^a epitope and cross reacts with simple Le^a and extended Le^a epitopes. NCCT-ST-421 induced ADCC and CDC to antigen-positive cells. It was shown to induce direct cell death as well through apoptotic mechanism. Although NCCT-ST-421 showed promising antitumour responses, no further details regarding clinical studies were described.

4.6. Direct killing anti-glycan monoclonal antibody

Oncosis is a progressive cell death process initially involving the impairment of ionic pumps of the cell membrane accompanied by cellular and organelle swelling. Subsequently, a gradually increase in membrane permeability due to an increasing cytosolic calcium concentration as well as rearrangement of cytoskeletal proteins results in pore formation in the cell membrane [130, 131]. It has been well accepted that tumours are able to manipulate the tumour microenvironment by releasing cytokines and other soluble factors, which create an immunosuppressive environment [132]. Release of cellular content into immunosuppressive tumour microenvironment via mAb-induced pore formation during oncosis may help in evoking immune responses via the release of danger-associated molecular patterns ('DAMPs'; [133]).

FG88 (FG88.2 and FG88.7) are internalising murine IgG3 mAbs recognise $Le^{a\text{-}c\text{-}x}$ glycans (**Figure 6**) overexpressed on a wide range of tumour cells and tissues at subnanomolar potency [134]. The FG88.2 mAb showed excellent tumour cell surface antigen binding and good levels of binding to a large percentage of tumours with low level binding to a limited number of normal tissues by immunohistochemistry. The significant association of strong FG88.2 binding with poor outcome in the colorectal sample cohort, independent of stage and vascular invasion, suggests that FG88.2 mAb has great potential targeting the most aggressive

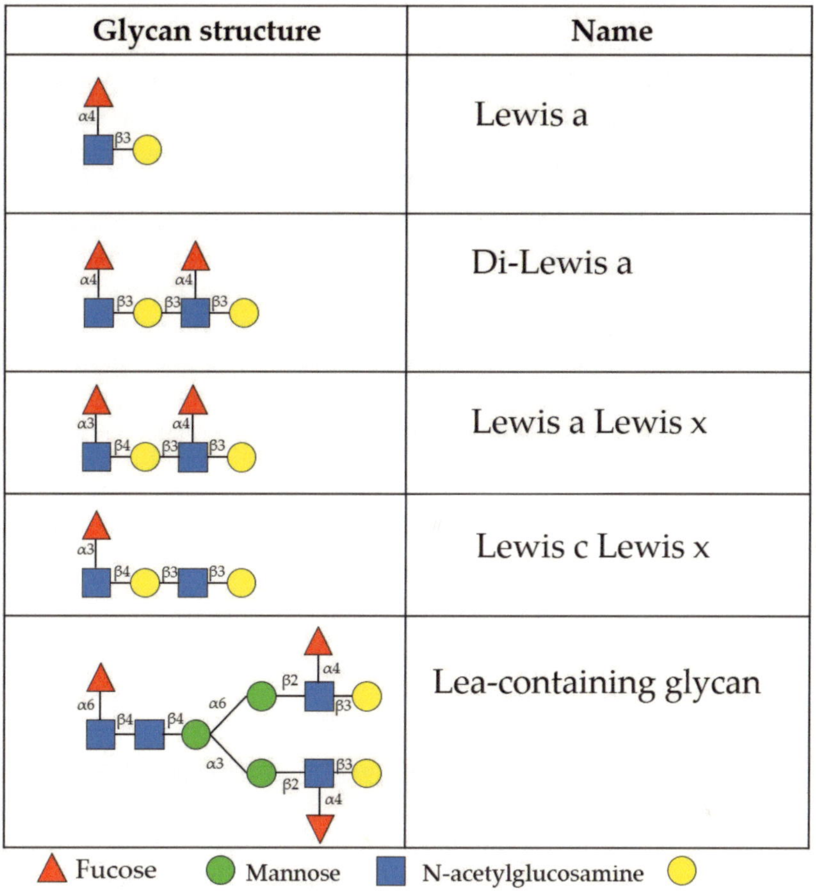

Figure 6. Details of glycan binding by FG88 mAbs.

colorectal cancers. FG88 mAbs induce potent ADCC, CDC and direct killing of tumour cells via oncosis. In the in vivo xenograft study, FG88 mAbs eradicated both primary and metastatic tumours. By releasing tumour cellular contents via mAb-induced pores, the FG88 mAbs might also be able to reverse the immunosuppressive tumour microenvironment leading to effective presentation of multiple epitopes from the lysed tumour cells and amplifying the antitumour immune response.

Several other anti-glycan mAbs can also induce similar direct tumour killing to FG88. MAb 84, when bound to human embryonic stem cells, induced cytoskeletal protein (α-actinin, paxillin and talin) degradation, which in turned increased the mobility of the plasma membrane, resulting in the clustering of antigens on the cell surface. Following antigen clustering, formation of pores through the plasma membrane resulted in oncosis [135]. N-glycolylneuraminic acid (NeuGc) is a sialic acid variant of N-acetylneuraminic acid (NeuAc). Humans cannot synthesise NeuGc because they lack the enzyme cytidine monophospho-N-acetylneuraminic acid hydroxylase (CMAH), responsible for its biosynthesis.

However, recent evidence suggests that NeuGc can be incorporated into human glycan from dietary sources. More interestingly, although not fully understood, human tumours actively incorporate NeuGc at a much higher rate than normal primary cells. As such, it is highly expressed in several human cancer cells [136], making it an appealing target for immunotherapy. It is noteworthy that natural circulating antibodies to NeuGc can be detected in normal human serum and that these antibodies have been shown to induce complement-dependent cell lysis [137]. Anti-NGcGM3 14 F7 mAb induced rapid cell death which was accompanied by cellular swelling, membrane lesion formation and cytoskeleton activation and cell aggregation. Moreover, no evidences of DNA fragmentation, chromatin modification or caspase activation were found. But 14 F7-treated cells showed large lesions at the plasma membrane, much bigger pores created by complement, perforin or bacterial toxins, suggesting an oncosis-like phenomenon [138]. Currently, the NeuGc-ganglioside anti-idiotypic mAb, Racotumomab (1E10), is under development. By molecular mimicry, a selected anti-idiotypic mAb will behave like the original antigen. In a phase III randomised trial in non-small-cell lung cancer, Racotumomab conferred a significant survival advantage [139] and induced anti-NeuGc antibodies capable of killing tumour cells by a mechanism similar to oncosis, validating the approach and highlighting the beneficial value of antibodies mediating oncosis for therapeutic purposes [140].

5. Conclusion

Glycoconjugates are major components of cells. They are involved in defining and modulating multiple key physiological processes in normal tissues. Aberrant glycosylation in cancer lead to the modification of glycosylations, resulted in the generation of TACAs, which drive several biological processes in cancer. Investigation of the molecular basis underlying these glycan modifications will aid in the understanding of cancer immunology as well as the development of anti-glycan therapeutic mAbs. Furthermore, rapid advances in glycomics and glycoproteomics will have a major impact on the unravelling of novel targets for cancer treatment.

Author details

Jia Xin Chua and Lindy Durrant*

*Address all correspondence to: lindy.durrant@nottingham.ac.uk

University of Nottingham, Nottingham, United Kingdom

References

[1] Tuccillo FM, de Laurentiis A, Palmieri C, Fiume G, Bonelli P, Borrelli A, et al. Aberrant glycosylation as biomarker for cancer: focus on CD43. BioMed Research International 2014;2014:742831.

[2] Fuster MM, Esko JD. The sweet and sour of cancer: glycans as novel therapeutic targets. Nature Reviews Cancer 2005;5:526–42.

[3] Padler-Karavani V. Aiming at the sweet side of cancer: aberrant glycosylation as possible target for personalized-medicine. Cancer Letters 2014;352:102–12.

[4] Spiro RG. Protein glycosylation: nature, distribution, enzymatic formation, and disease implications of glycopeptide bonds. Glycobiology 2002;12:43R–56R.

[5] Nagae M, Yamaguchi Y. Function and 3D structure of the N-glycans on glycoproteins. International Journal of Molecular Sciences 2012;13:8398–429.

[6] de Beer T, Vliegenthart JF, Loffler A, Hofsteenge J. The hexopyranosyl residue that is C-glycosidically linked to the side chain of tryptophan-7 in human RNase Us is alpha-mannopyranose. Biochemistry 1995;34:11785–9.

[7] Hofsteenge J, Muller DR, de Beer T, Loffler A, Richter WJ, Vliegenthart JF. New type of linkage between a carbohydrate and a protein: C-glycosylation of a specific tryptophan residue in human RNase Us. Biochemistry 1994;33:13524–30.

[8] Haynes PA. Phosphoglycosylation: a new structural class of glycosylation? Glycobiology 1998;8:1–5.

[9] Ferguson MA. The structure, biosynthesis and functions of glycosylphosphatidylinositol anchors, and the contributions of trypanosome research. Journal of Cell Science 1999;112(Pt 17):2799–809.

[10] Durrant LG, Noble P, Spendlove I. Immunology in the clinic review series; focus on cancer: glycolipids as targets for tumour immunotherapy. Clinical and Experimental Immunology 2012;167:206–15.

[11] Zhang D, Zhang G, Hayden MS, Greenblatt MB, Bussey C, Flavell RA, et al. A toll-like receptor that prevents infection by uropathogenic bacteria. Science 2004;303:1522–6.

[12] Lingwood CA. Glycosphingolipid functions. Cold Spring Harbor Perspectives in Biology 2011;2–3.

[13] Ichikawa S, Sakiyama H, Suzuki G, Hidari KI, Hirabayashi Y. Expression cloning of a cDNA for human ceramide glucosyltransferase that catalyzes the first glycosylation step of glycosphingolipid synthesis. Proceedings of the National Academy of Sciences of the United States of America 1996;93:4638–43.

[14] Futerman AH, Pagano RE. Determination of the intracellular sites and topology of glucosylceramide synthesis in rat liver. The Biochemical Journal 1991;280(Pt 2):295–302.

[15] Jeckel D, Karrenbauer A, Burger KN, van Meer G, Wieland F. Glucosylceramide is synthesized at the cytosolic surface of various Golgi subfractions. The Journal of Cell Biology 1992;117:259–67.

[16] Maccioni HJ, Giraudo CG, Daniotti JL. Understanding the stepwise synthesis of glycolipids. Neurochemical Research 2002;27:629–36.

[17] Hakomori S. Tumor-associated carbohydrate antigens defining tumor malignancy: basis for development of anti-cancer vaccines. Advances in Experimental Medicine and Biology 2001;491:369–402.

[18] Dube DH, Bertozzi CR. Glycans in cancer and inflammation – potential for therapeutics and diagnostics. Nature Reviews Drug Discovery 2005;4:477–88.

[19] Pochechueva T, Jacob F, Fedier A, Heinzelmann-Schwarz V. Tumor-associated glycans and their role in gynecological cancers: accelerating translational research by novel high-throughput approaches. Metabolites 2012;2:913–39.

[20] Dennis JW, Laferte S, Waghorne C, Breitman ML, Kerbel RS. Beta 1-6 branching of Asn-linked oligosaccharides is directly associated with metastasis. Science 1987;236:582–5.

[21] Kumamoto K, Goto Y, Sekikawa K, Takenoshita S, Ishida N, Kawakita M, et al. Increased expression of UDP-galactose transporter messenger RNA in human colon cancer tissues and its implication in synthesis of Thomsen-Friedenreich antigen and sialyl Lewis A/X determinants. Cancer Research 2001;61:4620–7.

[22] Kellokumpu S, Sormunen R, Kellokumpu I. Abnormal glycosylation and altered Golgi structure in colorectal cancer: dependence on intra-Golgi pH. FEBS Letters 2002;516:217–24.

[23] Chen L, Zhang W, Fregien N, Pierce M. The her-2/neu oncogene stimulates the transcription of N-acetylglucosaminyltransferase V and expression of its cell surface oligosaccharide products. Oncogene 1998;17:2087–93.

[24] Taniguchi N, Miyoshi E, Ko JH, Ikeda Y, Ihara Y. Implication of N-acetylglucosaminyltransferases III and V in cancer: gene regulation and signaling mechanism. Biochimica et Biophysica Acta 1999;1455:287–300.

[25] Miyoshi E, Terao M, Kamada Y. Physiological roles of N-acetylglucosaminyltransferase V(GnT-V) in mice. BMB Reports 2012;45:554–9.

[26] Ishida H, Togayachi A, Sakai T, Iwai T, Hiruma T, Sato T, et al. A novel beta1,3-N-acetylglucosaminyltransferase (beta3Gn-T8), which synthesizes poly-N-acetyllactosamine, is dramatically upregulated in colon cancer. FEBS Letters 2005;579:71–8.

[27] Mori S, Aoyagi Y, Yanagi M, Suzuki Y, Asakura H. Serum N-acetylglucosaminyltransferase III activities in hepatocellular carcinoma. Journal of Gastroenterology and Hepatology 1998;13:610–9.

[28] Song EY, Kang SK, Lee YC, Park YG, Chung TH, Kwon DH, et al. Expression of bisecting N-acetylglucosaminyltransferase-III in human hepatocarcinoma tissues, fetal liver tissues, and hepatoma cell lines of Hep3B and HepG2. Cancer Investigation 2001;19:799–807.

[29] Sewell R, Backstrom M, Dalziel M, Gschmeissner S, Karlsson H, Noll T, et al. The ST6GalNAc-I sialyltransferase localizes throughout the Golgi and is responsible for the synthesis of the tumor-associated sialyl-Tn O-glycan in human breast cancer. The Journal of Biological Chemistry 2006;281:3586–94.

[30] Julien S, Ivetic A, Grigoriadis A, QiZe D, Burford B, Sproviero D, et al. Selectin ligand sialyl-Lewis x antigen drives metastasis of hormone-dependent breast cancers. Cancer Research 2011;71:7683–93.

[31] Ogawa J, Inoue H, Koide S. Expression of alpha-1,3-fucosyltransferase type IV and VII genes is related to poor prognosis in lung cancer. Cancer Research 1996;56:325–9.

[32] Togayachi A, Kudo T, Ikehara Y, Iwasaki H, Nishihara S, Andoh T, et al. Up-regulation of Lewis enzyme (Fuc-TIII) and plasma-type alpha1,3fucosyltransferase (Fuc-TVI) expression determines the augmented expression of sialyl Lewis x antigen in non-small cell lung cancer. International Journal of Cancer 1999;83:70–9.

[33] Ju T, Wang Y, Aryal RP, Lehoux SD, Ding X, Kudelka MR, et al. Tn and sialyl-Tn antigens, aberrant O-glycomics as human disease markers. Proteomics Clinical Applications 2013;7:618–31.

[34] Terasawa K, Furumoto H, Kamada M, Aono T. Expression of Tn and sialyl-Tn antigens in the neoplastic transformation of uterine cervical epithelial cells. Cancer Research 1996;56:2229–32.

[35] Cao Y, Stosiek P, Springer GF, Karsten U. Thomsen-Friedenreich-related carbohydrate antigens in normal adult human tissues: a systematic and comparative study. Histochemistry and Cell Biology 1996;106:197–207.

[36] Springer GF. Immunoreactive T and Tn epitopes in cancer diagnosis, prognosis, and immunotherapy. Journal of Molecular Medicine (Berlin, Germany) 1997;75:594–602.

[37] Brockhausen I, Yang J, Dickinson N, Ogata S, Itzkowitz SH. Enzymatic basis for sialyl-Tn expression in human colon cancer cells. Glycoconjugate Journal 1998;15:595–603.

[38] Dalziel M, Whitehouse C, McFarlane I, Brockhausen I, Gschmeissner S, Schwientek T, et al. The relative activities of the C2GnT1 and ST3Gal-I glycosyltransferases determine O-glycan structure and expression of a tumor-associated epitope on MUC1. The Journal of Biological Chemistry 2001;276:11007–15.

[39] Marcos NT, Cruz A, Silva F, Almeida R, David L, Mandel U, et al. Polypeptide GalNAc-transferases, ST6GalNAc-transferase I, and ST3Gal-transferase I expression in gastric carcinoma cell lines. The Journal of Histochemistry and Cytochemistry: Official Journal of the Histochemistry Society 2003;51:761–71.

[40] Wagner KW, Punnoose EA, Januario T, Lawrence DA, Pitti RM, Lancaster K, et al. Death-receptor O-glycosylation controls tumor-cell sensitivity to the proapoptotic ligand Apo2L/TRAIL. Nature Medicine 2007;13:1070–7.

[41] Forrester JA, Ambrose EJ, Macpherson JA. Electrophoretic investigations of a clone of hamster fibroblasts and polyoma-transformed cells from the same population. Nature 1962;196:1068–70.

[42] Forrester JA, Ambrose EJ, Stoker MG. Microelectrophoresis of normal and transformed clones of hamster kidney fibroblasts. Nature 1964;201:945–6.

[43] Gasic G, Gasic T. Removal and regeneration of the cell coating in tumour cells. Nature 1962;196:170.

[44] Bull C, Stoel MA, den Brok MH, Adema GJ. Sialic acids sweeten a tumor's life. Cancer Research 2014;74:3199–204.

[45] Altevogt P, Fogel M, Cheingsong-Popov R, Dennis J, Robinson P, Schirrmacher V. Different patterns of lectin binding and cell surface sialylation detected on related high- and low-metastatic tumor lines. Cancer Research 1983;43:5138–44.

[46] Yogeeswaran G, Salk PL. Metastatic potential is positively correlated with cell surface sialylation of cultured murine tumor cell lines. Science 1981;212:1514–6.

[47] Schneider F, Kemmner W, Haensch W, Franke G, Gretschel S, Karsten U, et al. Overexpression of sialyltransferase CMP-sialic acid:Galbeta1,3GalNAc-R alpha6-Sialyltransferase is related to poor patient survival in human colorectal carcinomas. Cancer Research 2001;61:4605–11.

[48] Angata T, Varki A. Chemical diversity in the sialic acids and related alpha-keto acids: an evolutionary perspective. Chemical Reviews 2002;102:439–69.

[49] Pillai S, Netravali IA, Cariappa A, Mattoo H. Siglecs and immune regulation. Annual Review of Immunology 2012;30:357–92.

[50] Kim YJ, Varki A. Perspectives on the significance of altered glycosylation of glycoproteins in cancer. Glycoconjugate Journal 1997;14:569–76.

[51] Brockhausen I. Mucin-type O-glycans in human colon and breast cancer: glycodynamics and functions. EMBO Reports 2006;7:599–604.

[52] Hollingsworth MA, Swanson BJ. Mucins in cancer: protection and control of the cell surface. Nature Reviews Cancer 2004;4:45–60.

[53] Dall'Olio F, Chiricolo M. Sialyltransferases in cancer. Glycoconjugate Journal 2001;18:841–50.

[54] Gessner P, Riedl S, Quentmaier A, Kemmner W. Enhanced activity of CMP-neuAc:Gal beta 1-4GlcNAc:alpha 2,6-sialyltransferase in metastasizing human colorectal tumor tissue and serum of tumor patients. Cancer Letters 1993;75:143–9.

[55] Hedlund M, Ng E, Varki A, Varki NM. Alpha 2-6-Linked sialic acids on N-glycans modulate carcinoma differentiation in vivo. Cancer Research 2008;68:388–94.

[56] Wang PH, Lee WL, Lee YR, Juang CM, Chen YJ, Chao HT, et al. Enhanced expression of alpha 2,6-sialyltransferase ST6Gal I in cervical squamous cell carcinoma. Gynecologic Oncology 2003;89:395–401.

[57] Julien S, Adriaenssens E, Ottenberg K, Furlan A, Courtand G, Vercoutter-Edouart AS, et al. ST6GalNAc I expression in MDA-MB-231 breast cancer cells greatly modifies their O-glycosylation pattern and enhances their tumourigenicity. Glycobiology 2006;16:54–64.

[58] Alley WR, Jr., Novotny MV. Glycomic analysis of sialic acid linkages in glycans derived from blood serum glycoproteins. Journal of Proteome Research 2010;9:3062–72.

[59] Vasseur JA, Goetz JA, Alley WR, Jr., Novotny MV. Smoking and lung cancer-induced changes in N-glycosylation of blood serum proteins. Glycobiology 2012;22:1684–708.

[60] Saldova R, Fan Y, Fitzpatrick JM, Watson RW, Rudd PM. Core fucosylation and alpha2-3 sialylation in serum N-glycome is significantly increased in prostate cancer comparing to benign prostate hyperplasia. Glycobiology 2011;21:195–205.

[61] Yamamoto H, Saito T, Kaneko Y, Kersey D, Yong VW, Bremer EG, et al. Alpha2,3-sialyltransferase mRNA and alpha2,3-linked glycoprotein sialylation are increased in malignant gliomas. Brain Research 1997;755:175–9.

[62] Wang PH, Lee WL, Juang CM, Yang YH, Lo WH, Lai CR, et al. Altered mRNA expressions of sialyltransferases in ovarian cancers. Gynecologic Oncology 2005;99:631–9.

[63] Ravn V, Dabelsteen E. Tissue distribution of histo-blood group antigens. Acta Pathologica, Microbiologica, et Immunologica Scandinavica 2000;108:1–28.

[64] Yuriev E, Farrugia W, Scott AM, Ramsland PA. Three-dimensional structures of carbohydrate determinants of Lewis system antigens: implications for effective antibody targeting of cancer. Immunology and Cell Biology 2005;83:709–17.

[65] Krug LM, Milton DT, Jungbluth AA, Chen LC, Quaia E, Pandit-Taskar N, et al. Targeting Lewis Y (Le(y)) in small cell lung cancer with a humanized monoclonal antibody, hu3S193: a pilot trial testing two dose levels. Journal of Thoracic Oncology 2007;2:947–52.

[66] Soejima M, Koda Y. Molecular mechanisms of Lewis antigen expression. Legal Medicine (Tokyo, Japan) 2005;7:266–9.

[67] Ballare C, Barrio M, Portela P, Mordoh J. Functional properties of FC-2.15, a monoclonal antibody that mediates human complement cytotoxicity against breast cancer cells. Cancer Immunology, Immunotherapy 1995;41:15–22.

[68] Cazet A, Julien S, Bobowski M, Burchell J, Delannoy P. Tumour-associated carbohydrate antigens in breast cancer. Breast Cancer Research 2010;12:204.

[69] Kannagi R, Izawa M, Koike T, Miyazaki K, Kimura N. Carbohydrate-mediated cell adhesion in cancer metastasis and angiogenesis. Cancer Science 2004;95:377–84.

[70] Ragupathi G, Damani P, Srivastava G, Srivastava O, Sucheck SJ, Ichikawa Y, et al. Synthesis of sialyl Lewis(a) (sLe (a), CA19-9) and construction of an immunogenic sLe(a) vaccine. Cancer Immunology, Immunotherapy 2009;58:1397–405.

[71] Croce MV, Isla-Larrain M, Rabassa ME, Demichelis S, Colussi AG, Crespo M, et al. Lewis x is highly expressed in normal tissues: a comparative immunohistochemical study and literature revision. Pathology and Oncology Research 2007;13:130–8.

[72] Heimburg-Molinaro J, Lum M, Vijay G, Jain M, Almogren A, Rittenhouse-Olson K. Cancer vaccines and carbohydrate epitopes. Vaccine 2011;29:8802–26.

[73] Shimodaira K, Nakayama J, Nakamura N, Hasebe O, Katsuyama T, Fukuda M. Carcinoma-associated expression of core 2 beta-1,6-N-acetylglucosaminyltransferase gene in human colorectal cancer: role of O-glycans in tumor progression. Cancer Research 1997;57:5201–6.

[74] Ogawa J, Sano A, Inoue H, Koide S. Expression of Lewis-related antigen and prognosis in stage I non-small cell lung cancer. The Annals of Thoracic Surgery 1995;59:412–5.

[75] Birkle S, Zeng G, Gao L, Yu RK, Aubry J. Role of tumor-associated gangliosides in cancer progression. Biochimie 2003;85:455–63.

[76] Handa K, Hakomori SI. Carbohydrate to carbohydrate interaction in development process and cancer progression. Glycoconjugate Journal 2012;29:627–37.

[77] Lopez PH, Schnaar RL. Gangliosides in cell recognition and membrane protein regulation. Current Opinion in Structural Biology 2009;19:549–57.

[78] Fredman P, Hedberg K, Brezicka T. Gangliosides as therapeutic targets for cancer. BioDrugs 2003;17:155–67.

[79] Mujoo K, Cheresh DA, Yang HM, Reisfeld RA. Disialoganglioside GD2 on human neuroblastoma cells: target antigen for monoclonal antibody-mediated cytolysis and suppression of tumor growth. Cancer Research 1987;47:1098–104.

[80] Grant SC, Kostakoglu L, Kris MG, Yeh SD, Larson SM, Finn RD, et al. Targeting of small-cell lung cancer using the anti-GD2 ganglioside monoclonal antibody 3 F8: a pilot trial. European Journal of Nuclear Medicine 1996;23:145–9.

[81] Zhang S, Cordon-Cardo C, Zhang HS, Reuter VE, Adluri S, Hamilton WB, et al. Selection of tumor antigens as targets for immune attack using immunohistochemistry: I. Focus on gangliosides. International Journal of Cancer 1997;73:42–9.

[82] Livingston PO, Hood C, Krug LM, Warren N, Kris MG, Brezicka T, et al. Selection of GM2, fucosyl GM1, globo H and polysialic acid as targets on small cell lung cancers for antibody mediated immunotherapy. Cancer Immunology, Immunotherapy 2005;54:1018–25.

[83] Cheresh DA, Pytela R, Pierschbacher MD, Klier FG, Ruoslahti E, Reisfeld RA. An Arg-Gly-Asp-directed receptor on the surface of human melanoma cells exists in an divalent cation-dependent functional complex with the disialoganglioside GD2. The Journal of Cell Biology 1987;105:1163–73.

[84] Wang X, Sun P, Al-Qamari A, Tai T, Kawashima I, Paller AS. Carbohydrate-carbohydrate binding of ganglioside to integrin alpha(5) modulates alpha(5)beta(1) function. The Journal of Biological Chemistry 2001;276:8436–44.

[85] Crocker PR, Clark EA, Filbin M, Gordon S, Jones Y, Kehrl JH, et al. Siglecs: a family of sialic-acid binding lectins. Glycobiology 1998;8:v.

[86] Ito A, Handa K, Withers DA, Satoh M, Hakomori S. Binding specificity of siglec7 to disialogangliosides of renal cell carcinoma: possible role of disialogangliosides in tumor progression. FEBS Letters 2001;504:82–6.

[87] Khatib AM, Kontogiannea M, Fallavollita L, Jamison B, Meterissian S, Brodt P. Rapid induction of cytokine and E-selectin expression in the liver in response to metastatic tumor cells. Cancer Research 1999;59:1356–61.

[88] Kojima N, Shiota M, Sadahira Y, Handa K, Hakomori S. Cell adhesion in a dynamic flow system as compared to static system. Glycosphingolipid-glycosphingolipid interaction in the dynamic system predominates over lectin- or integrin-based mechanisms in adhesion of B16 melanoma cells to non-activated endothelial cells. The Journal of Biological Chemistry 1992;267:17264–70.

[89] Hakomori S. Tumor malignancy defined by aberrant glycosylation and sphingo(glyco) lipid metabolism. Cancer Research 1996;56:5309–18.

[90] Biswas K, Richmond A, Rayman P, Biswas S, Thornton M, Sa G, et al. GM2 expression in renal cell carcinoma: potential role in tumor-induced T-cell dysfunction. Cancer Research 2006;66:6816–25.

[91] Biswas S, Biswas K, Richmond A, Ko J, Ghosh S, Simmons M, et al. Elevated levels of select gangliosides in T cells from renal cell carcinoma patients is associated with T cell dysfunction. Journal of Immunology 2009;183:5050–8.

[92] Chu JW, Sharom FJ. Gangliosides inhibit T-lymphocyte proliferation by preventing the interaction of interleukin-2 with its cell surface receptors. Immunology 1993;79:10–7.

[93] Irani DN, Lin KI, Griffin DE. Brain-derived gangliosides regulate the cytokine production and proliferation of activated T cells. The Journal of Immunology 1996;157:4333–40.

[94] Morioka N, Furue M, Tsuchida T, Ishibashi Y. Gangliosides inhibit the proliferation of human T cells stimulated with interleukin-4 or interleukin-2. The Journal of Dermatology 1991;18:447–53.

[95] Heitger A, Ladisch S. Gangliosides block antigen presentation by human monocytes. Biochimica et Biophysica Acta 1996;1303:161–8.

[96] Uzzo RG, Rayman P, Kolenko V, Clark PE, Cathcart MK, Bloom T, et al. Renal cell carcinoma-derived gangliosides suppress nuclear factor-kappaB activation in T cells. The Journal of Clinical Investigation 1999;104:769–76.

[97] Thornton MV, Kudo D, Rayman P, Horton C, Molto L, Cathcart MK, et al. Degradation of NF-kappa B in T cells by gangliosides expressed on renal cell carcinomas. The Journal of Immunology 2004;172:3480–90.

[98] Gonzalez SA. Novel biomarkers for hepatocellular carcinoma surveillance: has the future arrived? Hepatobiliary Surgery and Nutrition 2014;3:410–4.

[99] Kirwan A, Utratna M, O'Dwyer ME, Joshi L, Kilcoyne M. Glycosylation-based serum biomarkers for cancer diagnostics and prognostics. BioMed Research International 2015;2015:490531.

[100] Bergstrand CG, Czar B. Demonstration of a new protein fraction in serum from the human fetus. Scandinavian Journal of Clinical and Laboratory Investigation 1956;8:174.

[101] Kobayashi M, Kuroiwa T, Suda T, Tamura Y, Kawai H, Igarashi M, et al. Fucosylated fraction of alpha-fetoprotein, L3, as a useful prognostic factor in patients with hepatocellular carcinoma with special reference to low concentrations of serum alpha-fetoprotein. Hepatology Research: The Official Journal of the Japan Society of Hepatology 2007;37:914–22.

[102] Johnson PJ, Poon TC, Hjelm NM, Ho CS, Blake C, Ho SK. Structures of disease-specific serum alpha-fetoprotein isoforms. British Journal of Cancer 2000;83:1330–7.

[103] Ito Y, Miyauchi A, Yoshida H, Uruno T, Nakano K, Takamura Y, et al. Expression of alpha1,6-fucosyltransferase (FUT8) in papillary carcinoma of the thyroid: its linkage to biological aggressiveness and anaplastic transformation. Cancer Letters 2003;200:167–72.

[104] Osumi D, Takahashi M, Miyoshi E, Yokoe S, Lee SH, Noda K, et al. Core fucosylation of E-cadherin enhances cell-cell adhesion in human colon carcinoma WiDr cells. Cancer Science 2009;100:888–95.

[105] Kyselova Z, Mechref Y, Al Bataineh MM, Dobrolecki LE, Hickey RJ, Vinson J, et al. Alterations in the serum glycome due to metastatic prostate cancer. Journal of Proteome Research 2007;6:1822–32.

[106] Peracaula R, Barrabes S, Sarrats A, Rudd PM, de Llorens R. Altered glycosylation in tumours focused to cancer diagnosis. Disease Markers 2008;25:207–18.

[107] Ohyama C, Hosono M, Nitta K, Oh-eda M, Yoshikawa K, Habuchi T, et al. Carbohydrate structure and differential binding of prostate specific antigen to Maackia amurensis lectin between prostate cancer and benign prostate hypertrophy. Glycobiology 2004;14:671–9.

[108] Peracaula R, Tabares G, Royle L, Harvey DJ, Dwek RA, Rudd PM, et al. Altered glycosylation pattern allows the distinction between prostate-specific antigen (PSA) from normal and tumor origins. Glycobiology 2003;13:457–70.

[109] Tajiri M, Ohyama C, Wada Y. Oligosaccharide profiles of the prostate specific antigen in free and complexed forms from the prostate cancer patient serum and in seminal plasma: a glycopeptide approach. Glycobiology 2008;18:2–8.

[110] Sarrats A, Comet J, Tabares G, Ramirez M, Aleixandre RN, de Llorens R, et al. Differential percentage of serum prostate-specific antigen subforms suggests a new way to improve prostate cancer diagnosis. The Prostate 2010;70:1–9.

[111] Isono T, Tanaka T, Kageyama S, Yoshiki T. Structural diversity of cancer-related and non-cancer-related prostate-specific antigen. Clinical Chemistry 2002;48:2187–94.

[112] Kannagi R. Carbohydrate antigen sialyl Lewis a – its pathophysiological significance and induction mechanism in cancer progression. Chang Gung Medical Journal 2007;30:189–209.

[113] Koprowski H, Steplewski Z, Mitchell K, Herlyn M, Herlyn D, Fuhrer P. Colorectal carcinoma antigens detected by hybridoma antibodies. Somatic Cell Genetics 1979;5:957–71.

[114] Ballehaninna UK, Chamberlain RS. The clinical utility of serum CA 19-9 in the diagnosis, prognosis and management of pancreatic adenocarcinoma: an evidence based appraisal. Journal of Gastrointestinal Oncology 2012;3:105–19.

[115] Duraker N, Hot S, Polat Y, Hobek A, Gencler N, Urhan N. CEA, CA 19-9, and CA 125 in the differential diagnosis of benign and malignant pancreatic diseases with or without jaundice. Journal of Surgical Oncology 2007;95:142–7.

[116] Liao Q, Zhao YP, Yang YC, Li LJ, Long X, Han SM. Combined detection of serum tumor markers for differential diagnosis of solid lesions located at the pancreatic head. Hepatobiliary & Pancreatic Diseases International 2007;6:641–5.

[117] Safi F, Roscher R, Bittner R, Schenkluhn B, Dopfer HP, Beger HG. High sensitivity and specificity of CA 19-9 for pancreatic carcinoma in comparison to chronic pancreatitis. Serological and immunohistochemical findings. Pancreas 1987;2:398–403.

[118] Vestergaard EM, Hein HO, Meyer H, Grunnet N, Jorgensen J, Wolf H, et al. Reference values and biological variation for tumor marker CA 19-9 in serum for different Lewis and secretor genotypes and evaluation of secretor and Lewis genotyping in a Caucasian population. Clinical Chemistry 1999;45:54–61.

[119] Weiner LM, Murray JC, Shuptrine CW. Antibody-based immunotherapy of cancer. Cell 2012;148:1081–4.

[120] Cheung NK, Saarinen UM, Neely JE, Landmeier B, Donovan D, Coccia PF. Monoclonal antibodies to a glycolipid antigen on human neuroblastoma cells. Cancer Research 1985;45:2642–9.

[121] Svennerholm L, Bostrom K, Fredman P, Jungbjer B, Lekman A, Mansson JE, et al. Gangliosides and allied glycosphingolipids in human peripheral nerve and spinal cord. Biochimica et Biophysica Acta 1994;1214:115–23.

[122] Yu AL, Gilman AL, Ozkaynak MF, London WB, Kreissman SG, Chen HX, et al. Anti-GD2 antibody with GM-CSF, interleukin-2, and isotretinoin for neuroblastoma. The New England Journal of Medicine 2010;363:1324–34.

[123] Hellstrom I, Garrigues HJ, Garrigues U, Hellstrom KE. Highly tumor-reactive, internalizing, mouse monoclonal antibodies to Le(y)-related cell surface antigens. Cancer Research 1990;50:2183–90.

[124] Tolcher AW, Sugarman S, Gelmon KA, Cohen R, Saleh M, Isaacs C, et al. Randomized phase II study of BR96-doxorubicin conjugate in patients with metastatic breast cancer. Journal of Clinical Oncology: Official Journal of the American Society of Clinical Oncology 1999;17:478–84.

[125] Ross HJ, Hart LL, Swanson PM, Rarick MU, Figlin RA, Jacobs AD, et al. A randomized, multicenter study to determine the safety and efficacy of the immunoconjugate

SGN-15 plus docetaxel for the treatment of non-small cell lung carcinoma. Lung Cancer 2006;54:69–77.

[126] Scott AM, Geleick D, Rubira M, Clarke K, Nice EC, Smyth FE, et al. Construction, production, and characterization of humanized anti-Lewis Y monoclonal antibody 3S193 for targeted immunotherapy of solid tumors. Cancer Research 2000;60:3254–61.

[127] Farrugia W, Scott AM, Ramsland PA. A possible role for metallic ions in the carbohydrate cluster recognition displayed by a Lewis Y specific antibody. PLoS One 2009;4:e7777.

[128] Shitara K, Hanai N, Kusano A, Furuya A, Yoshida H, Wada K, et al. Application of anti-sialyl Lea monoclonal antibody, KM231, for immunotherapy of cancer. Anticancer Research 1991;11:2003–13.

[129] Sawada R, Sun SM, Wu X, Hong F, Ragupathi G, Livingston PO, et al. Human monoclonal antibodies to sialyl-Lewis (CA19.9) with potent CDC, ADCC, and antitumor activity. Clinical Cancer Research: An Official Journal of the American Association for Cancer Research 2011;17:1024–32.

[130] Weerasinghe P, Buja LM. Oncosis: an important non-apoptotic mode of cell death. Experimental and Molecular Pathology 2012;93:302–8.

[131] Weerasinghe P, Hallock S, Brown RE, Loose DS, Buja LM. A model for cardiomyocyte cell death: insights into mechanisms of oncosis. Experimental and Molecular Pathology 2013;94:289–300.

[132] Ostman A, Augsten M. Cancer-associated fibroblasts and tumor growth – bystanders turning into key players. Current Opinion in Genetics & Development 2009;19:67–73.

[133] Krysko O, Love Aaes T, Bachert C, Vandenabeele P, Krysko DV. Many faces of DAMPs in cancer therapy. Cell Death and Disease 2013;4:e631.

[134] Chua JX, Vankemmelbeke M, McIntosh RS, Clarke PA, Moss R, Parsons T, et al. Monoclonal antibodies targeting leclex-related glycans with potent antitumor activity. Clinical Cancer Research: An Official Journal of the American Association for Cancer Research 2015;21:2963–74.

[135] Tan HL, Fong WJ, Lee EH, Yap M, Choo A. mAb 84, a cytotoxic antibody that kills undifferentiated human embryonic stem cells via oncosis. Stem Cells 2009;27:1792–801.

[136] Malykh YN, Schauer R, Shaw L. N-Glycolylneuraminic acid in human tumours. Biochimie 2001;83:623–34.

[137] Nguyen DH, Tangvoranuntakul P, Varki A. Effects of natural human antibodies against a nonhuman sialic acid that metabolically incorporates into activated and malignant immune cells. The Journal of Immunology 2005;175:228–36.

[138] Roque-Navarro L, Chakrabandhu K, de Leon J, Rodriguez S, Toledo C, Carr A, et al. Anti-ganglioside antibody-induced tumor cell death by loss of membrane integrity. Molecular Cancer Therapeutics 2008;7:2033–41.

[139] Alfonso S, Valdes-Zayas A, Santiesteban ER, Flores YI, Areces F, Hernandez M, et al. A randomized, multicenter, placebo-controlled clinical trial of racotumomab-alum vaccine as switch maintenance therapy in advanced non-small cell lung cancer patients. Clinical Cancer Research: An Official Journal of the American Association for Cancer Research 2014;20:3660–71.

[140] Hernandez AM, Rodriguez N, Gonzalez JE, Reyes E, Rondon T, Grinan T, et al. Anti-NeuGcGM3 antibodies, actively elicited by idiotypic vaccination in nonsmall cell lung cancer patients, induce tumor cell death by an oncosis-like mechanism. The Journal of Immunology 2011;186:3735–44.

Chapter 4

Self-Assembled Monolayers of Carbohydrate Derivatives on Gold Surfaces

Jay K. Bhattarai, Dharmendra Neupane,
Vasilii Mikhaylov, Alexei V. Demchenko and
Keith J. Stine

Additional information is available at the end of the chapter

http://dx.doi.org/10.5772/66194

Abstract

Self-assembled monolayers (SAMs) presenting carbohydrates (glycans) have been widely prepared on gold surfaces to mimic the carbohydrate surfaces that are involved in molecular recognition phenomena in living cells. The binding affinity of carbohydrate immbolized on SAM surfaces to various carbohydrate-binding proteins (such as lectins) can be studied by optical, electrochemical, piezoelectrical and thermal sensing techniques. The lectins present on the surface of pathogens (e.g., bacteria or viruses) can be used as targets for capturing onto carbohydrates immobilized on SAM surfaces. The immobilized carbohydrates can also be used for detecting different types of disease biomarkers present in bodily fluids. Synergistic properties of carbohydrate SAMs and gold nanoparticles can be used for vaccine preparation and drug delivery. By studying different types of glycans, their properties, and the behavior toward recognition of specific pathogens and biomarkers, we can develop not only new therapeutics but also enhance the diagnostic strategies of various diseases. In this chapter, we discuss carbohydrate-terminated SAMs and their common preparation strategies. Next, we focus on roles of different components of SAMs, characterization techniques, and applications.

Keywords: self-assembled monolayers (SAMs), carbohydrates, gold surface, click reaction, biosensing, carbohydrate-lectin interaction, *E-coli* detection

1. Introduction

Carbohydrates are biological molecules, present widely in nature in diverse forms and have varieties of functions [1]. Their role in living organisms is indispensable, whether it is as a structural support (e.g., cellulose and chitin), in energy storage (e.g., glycogen and starch), in the immune system, or for fertilization and development [2, 3]. The glycans are the carbohydrate parts of glycoconjugates, such as glycoproteins, glycolipids, or proteoglycans, and their importance in human health and disease is an ever expanding field [3]. A major part of the field concerns the study of the organization of carbohydrates at interfaces and their interaction with carbohydrate-binding proteins. Self-assembled monolayers (SAMs) of carbohydrate-terminated alkanethiols and of other carbohydrate derivatives conjugated to species pre-immobilized on gold such as branched polymers and dendrimers have served as model systems in many studies of these interactions and in the development of biosensors based on carbohydrate recognition [4, 5]. These SAMs have been formed both by direct immobilization of carbohydrate-terminated alkanethiols and by conjugation of glycans to pre-formed SAMs with reactive terminal groups [6]. The complexity of the carbohydrates immobilized range from monosaccharides to complex oligosaccharides of varied biological functions. Central to these studies is the goal of understanding the structure and organization of the SAMs, and this has been approached using a range of methods, including surface analysis, surface spectroscopy, scanning probe microscopy, and electrochemical methods. The binding of proteins to these SAMs has been followed using methods, including surface plasmon resonance (SPR), impedance spectroscopy, and quartz crystal microbalance (QCM). In this chapter, we will seek to review the literature concerning SAMs containing terminal carbohydrates, their fabrication by direct or indirect coupling methods, and their structural characterization. The applications of these SAMs in protein-binding studies and biosensor development will also be discussed.

2. Self-assembled monolayers

Organic molecules having functional head-groups (e.g., thiols, disulfides, and amines) and tail groups at the end of hydrophobic chains (e.g., alkanes and polyethylene glycols) can easily self-assemble on noble metal surfaces lowering the free energy at the interface to form densely packed monolayer films, called self-assembled monolayers (SAMs) [7]. Different types of functional groups can be attached to the terminal end of the hydrophobic part depending on the nature of the study, through which further chemistry can be performed linking fields of the material chemistry and organic/biochemistry. SAMs of organosulfur compounds are the most-studied SAMs to date because of strong thiol-gold bond formation [8, 9]. A schematic diagram of an ideal alkanethiol SAM immobilized on gold surfaces having terminal functional groups is shown in **Figure 1,** and the chemisorption reaction between thiol and gold is shown in Eq. (1) [10].

$$\text{R–SH} + \text{Au}_n^0 \rightarrow \text{R–S}^{(-)}\text{–Au}^{(+)} \cdot \text{Au}_n^0 + 1/2\text{H}_2 \tag{1}$$

One of the common methods of preparing SAMs of organosulfur derivatives is by immersing a metal substrate into dilute (1–10 mM) ethanolic solution of the desired organosulfur compound for 12–24 h under ambient conditions [11]. When sulfur atoms come in contact with a clean metal surface, they start forming monolayers instantly; however, the molecules reorganize themselves if left in the solution over a longer period, minimizing the density of defects [12]. Alkyl chains of SAMs arrange themselves in trans-conformation with nearly 20–30° tilt from normal to the metal surface [9]. However, studies have shown that overall arrangement

http://dx.doi.org/10.5772/66194

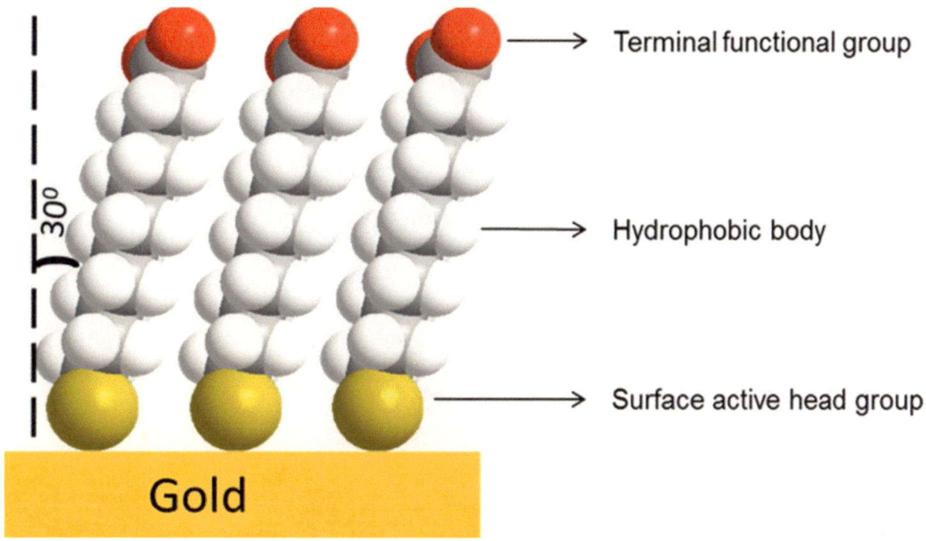

Figure 1. Schematic depiction of an ideal SAM formed on a gold substrate.

and binding of SAMs on gold surfaces depend on numbers of factors, including length of alkyl chains, the nature and distance between terminal functional groups, concentration and purity of adsorbate, immersion time, and substrate morphology [11].

3. Preparation of SAMs having terminal carbohydrates

Preparation of microarrays of carbohydrates to mimic the cell surface for the in vitro study of their interactions with pathogens or other biological molecules is very important. Microarrays present a surface onto which pathogens can undergo multivalent attachment amplifying the relative affinities as on cell surfaces and above that of a single ligand. The pathogens or biological molecules captured on the array can also be easily harvested and further tested. There are different methods to prepare arrays of carbohydrates on solid surfaces; one of the popular methods is through SAMs formation. Here, we will be specifically focusing on the SAM of carbohydrates prepared on a gold substrate through organosulfur molecules. The two common approaches (**Figure 2**) for forming SAMs of carbohydrates on gold surfaces are, (1) direct method: The carbohydrate molecules of interest are modified with the organosulfur molecule first, followed by direct SAMs formation on the gold surface; and (2) indirect method: The SAM of organosulfur molecule having a suitable terminal functional group is prepared on the gold surface first, followed by a reaction to conjugate it to the carbohydrate of interest.

3.1. Indirect methods

The indirect method of forming carbohydrate SAMs does not require preparation of organosulfur molecules already linked to the carbohydrate of interest. The strategy also avoids

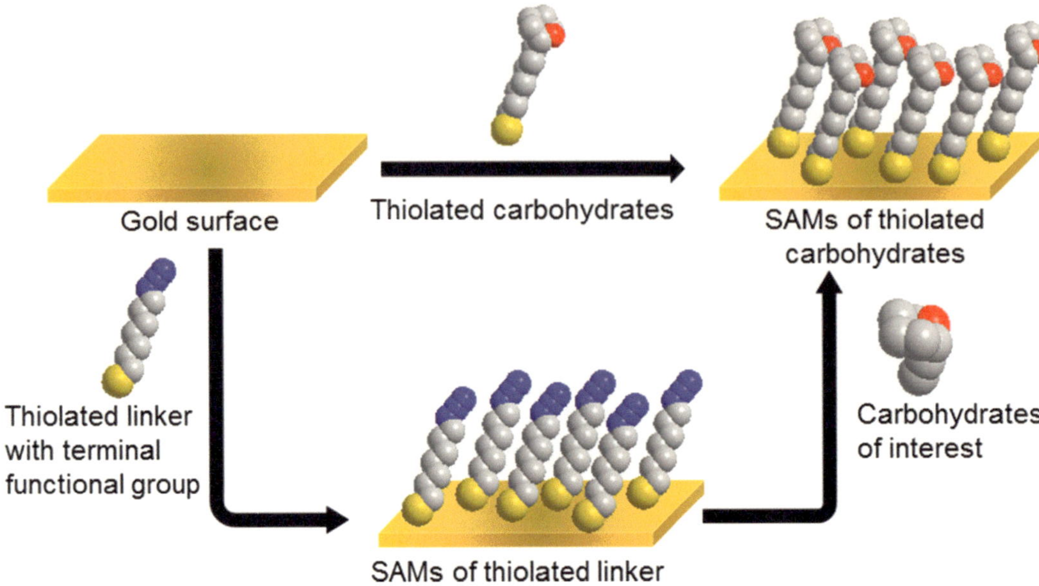

Figure 2. Schematic of two different strategies for forming the SAM of carbohydrates on gold surfaces.

the possibility that the prepared molecules may decompose or oxidize by the time they are used and might not form stable SAMs. An advantage of the indirect SAM formation method is that each step can be tracked in situ [13], which can be imagined as building a tower by stacking bricks on top of one another with cement between the bricks, whereas the direct method is like transferring a whole tower in one piece. Use of indirect methods for the preparation of carbohydrate SAMs date back to the 1990s. In 1995, Lofas used a step-by-step method to form SAMs containing dextran, a hydrophilic linear polymer based on 1,6-linked glucose units, on a gold surface [14]. As a first step, a SAM of 16-mercaptohexadecan-1-ol was formed on a gold surface, followed by the reaction of the exposed hydroxyl groups with epichlorohydrin to prepare terminal epoxy groups. The epoxy group was found to capture the dextran randomly under basic conditions. The immobilized dextran was subsequently reacted with bromoacetic acid to introduce terminal functional carboxylic groups, which were in turn activated using a mixture of N-ethyl-N'-dimethylaminopropyl-carbodiimide and N-hydroxysuccinimide to capture monoclonal antibodies (MAbs). Three different types of MAbs were then tested against their antigen, HIV protein p24. The as-prepared sensor was sensitive enough to distinguish differences in the affinity of the three MAbs toward p24 [14]. During the two decades since, different types of strategies have been discovered for preparing SAMs of carbohydrates using indirect methods, applicable for immobilizing both simple and complex carbohydrate molecules.

Click chemistry-based reactions are popular indirect methods for formation of carbohydrate SAMs. Click reactions skip the tremendous synthetic efforts required for the preparation of thiolated carbohydrates to be used for direct SAM formation. In addition, this reaction can tolerate a wide variety of functional groups and can be performed over a broad range of temperature and pH with minimal by-product formation [15, 16].

One of the early uses of click reaction for the carbohydrate SAM formation on a gold surface was performed by Houseman and Mrksich in 2002. They used Diels-Alder reaction to connect benzoquinone-terminated SAM surfaces to different common monosaccharides derivatized with cyclopentadiene and prepared carbochips [17]. The carbochips were then utilized for profiling lectin-binding specificity to their corresponding monosaccharides using SPR and confocal fluorescence microscopy as shown in **Figure 3**. The same group soon reported another click reaction strategy based on maleimide-thiol chemistry for preparing SAMs of carbohydrates [18]. They synthesized four different monosaccharides (mannose, galactose, glucose, and N-acetylglucosamine) having thiol groups at the anomeric centers, which reacted selectively with the terminal maleimide groups of pre-immobilized SAMs on gold surfaces to present carbohydrate-terminated SAMs. They then prepared a carbohydrate array to study the specificity of these carbohydrates to their corresponding lectins using confocal fluorescence microscopy, as before. This strategy of maleimide-thiol reaction can be used for preparing carbohydrate array SAMs having terminal mannoses and interrogate GFP-transfected *E coli* bacteria creating a highly biorepulsive linker [19].

The most commonly used click reaction for preparing carbohydrates SAMs, however, is the Cu (I)-catalyzed Huisgen 1, 3-dipolar cycloaddition or the copper(I)-catalyzed azide-alkyne cycloaddition (CuAAC) reaction [15]. The first step in this method is to form an organosulfur SAM having either alkyne or azide terminal groups [20, 21]. If the alkyne group is presented at the SAM's surface, it reacts with azide groups available on anomeric positions of carbohydrate molecules of interest and vice versa [22]. This click reaction results in formation of a 1,2,3-triazole ring with the SAM presenting the terminal carbohydrate [23]. Zhang and coworkers used this chemistry to immobilize azido sugars—mannose, lactose, and α-Gal trisaccharides—and studied their interactions with their corresponding specific lectins [24]. Using QCM, the apparent affinity constant (K_a) for the interaction of mannose–Con A, lactose–*Erythrina cristagalli* lectin, and α-Gal trisaccharides–polyclonal anti-Gal antibody were found to be $(8.7 \pm 2.8) \times 10^5$, $(4.6 \pm 2.4) \times 10^6$ and $(6.7 \pm 3.3) \times 10^6$ M^{-1}, respectively. The data were further supported with SPR, AFM, and electrochemical experiments. The prepared sensor was found to be very selective having negligible nonspecific adsorption. Later, the Tamiya laboratory used this strategy to electrochemically detect Alzheimer's related peptide amyloid-β by immobilizing sialic acid derivatives on gold nanoparticles deposited on a carbon electrode [21]. On capturing the peptides Aβ(1-40) and Aβ(1-42), a characteristic oxidation peak was observed at 0.6 V (vs. Ag/AgCl) through differential pulse voltammetry, which was further confirmed by AFM images showing increase in roughness on the surface after capturing of peptides by the sialic acid. In another study, using circular dichroism spectroscopy and fluorescence microscopy, SAM of sialic acid and 6-sulfo-GlcNAc was found to have tendency to change the conformations of Aβ(1–42) into β-sheet while also aggregating Aβ into fibrils having a larger diameter compare to that created by SAM of β-Glc [25]. In the same work, binding affinities of different common monosaccharide SAMs to their corresponding lectins were also studied and were found to be the range of 10^{-7}–10^{-8} M.

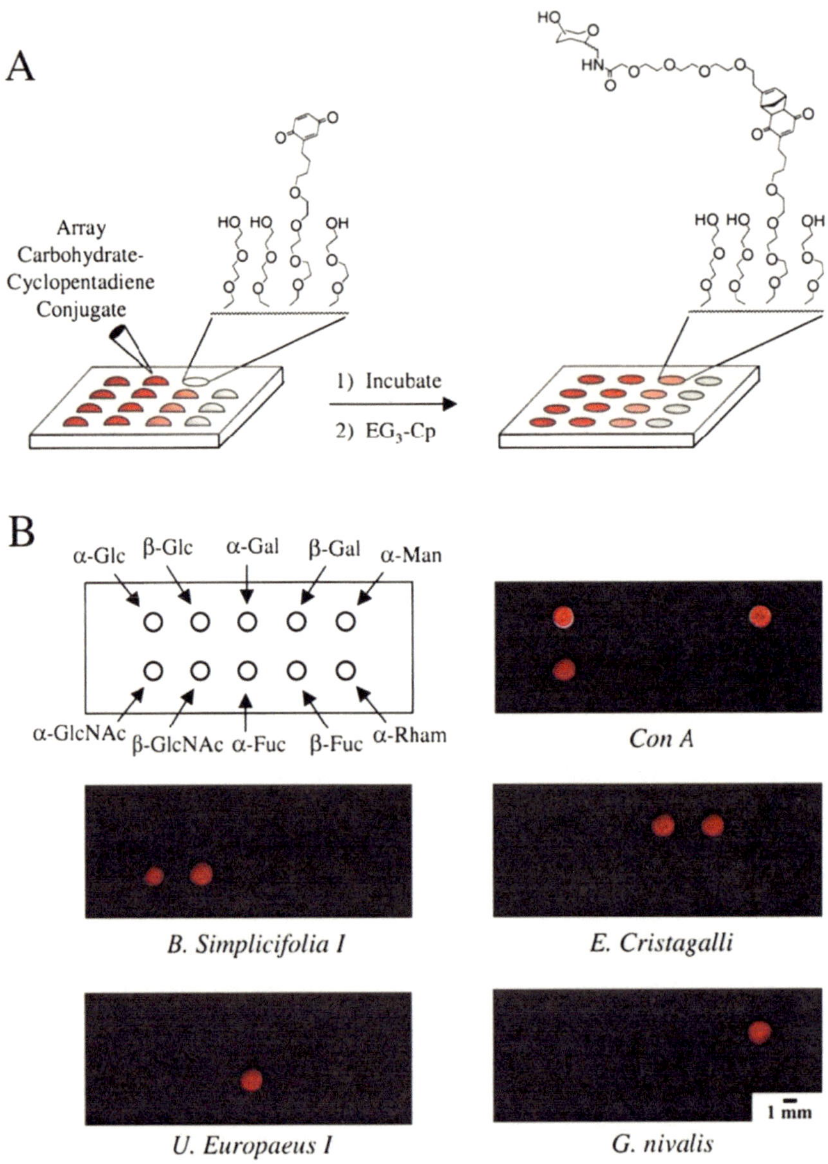

Figure 3. Profiling lectin binding specificities with a carbohydrate array (A). Carbohydrate arrays are prepared by spotting solutions of carbohydrate diene conjugates onto discrete regions of a monolayer presenting benzoquinone groups. After the reaction is complete, benzoquinone groups in the remaining regions of the monolayer can be inactivated by treatment with tri(ethylene glycol)-cyclopentadiene conjugate (EG3-Cp). (B) Identical carbohydrate chips were separately incubated with each of five rhodaminelabeled lectins (2 μM in DPBS) for 30 min, gently rinsed, and evaluated by confocal fluorescence microscopy. Fluorescent images of the resulting arrays are shown for each lectin. These images reveal that the proteins associate specifically with their known carbohydrate ligands on the array. Reprinted with permission from Ref. [17], Copyright 2002, Elsevier.

To create a well-defined multivalency, SAMs of glycodendrimers of various generations were prepared using CuAAC reactions on a gold surface (**Figure 4**) and characterized using

http://dx.doi.org/10.5772/66194

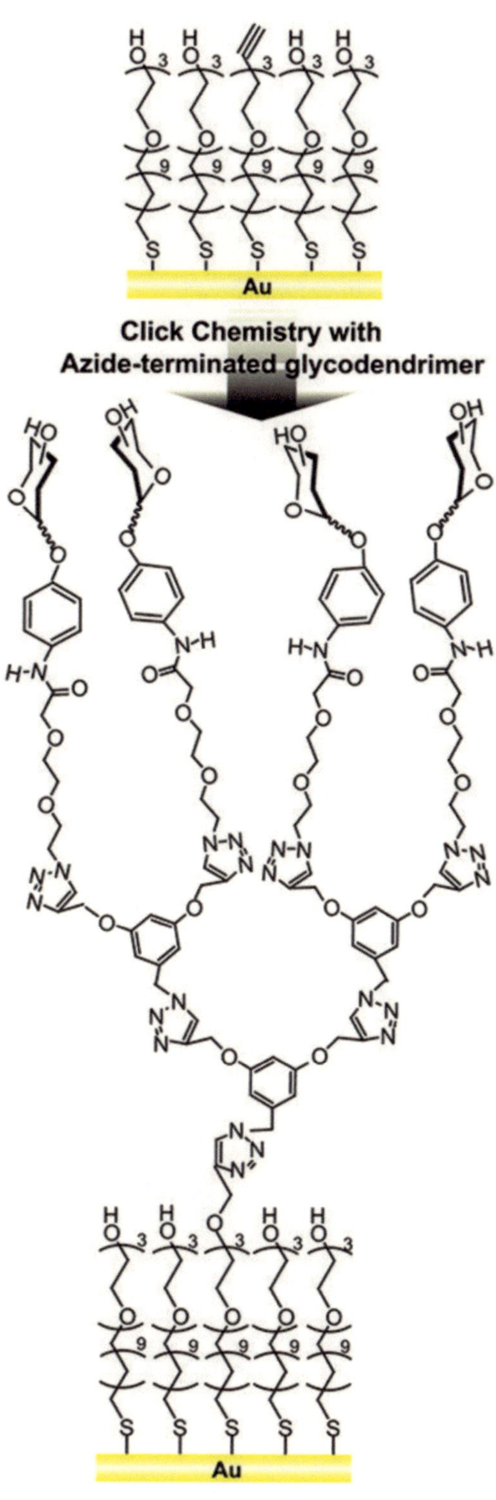

Figure 4. Schematic illustration for the process of glycodendrimer immobilization. Reprinted with permission from Ref. [4], Copyright 2009, Elsevier.

XPS, ellipsometry, MALDI-ToF mass spectrometry, cyclic voltammetry, ^1H and ^{13}C NMR, contact angle goniometry and FTIR [4, 26]. The equilibrium association constants (K_a) and kinetic association rate constants (k_{on}) and dissociation rate constants (k_{off}) for the interactions of terminal monosaccharides (α-Man, β-GlcNAc, and β-Gal) to their corresponding lectins were determined using SPR [4]. Similarly, dendritic mannosylated surfaces were used to enhanced recognition of *E. coli* bacteria with generation one dendrimer detecting 3570

Pre-formed SAM	**Functionalized carbohydrates**	**Reactive functional groups**
		Alkyne/Azide [4, 13, 21, 24, 25, 27]
		Azide/Alkyne [20]
		Amine/Isothiocyanate [5, 27]
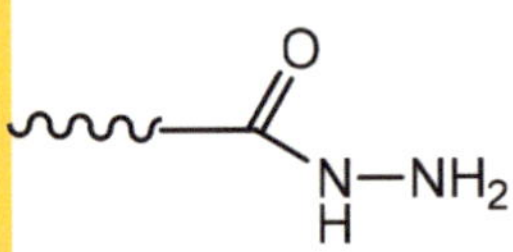		Hydrazide/unmodified carbohydrates [28]
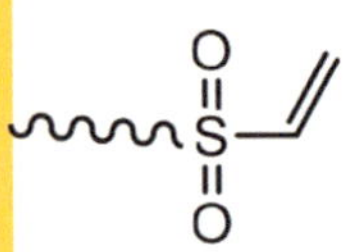	 R = SH, NH_2	Vinyl sulfone/amine or thiol [29]
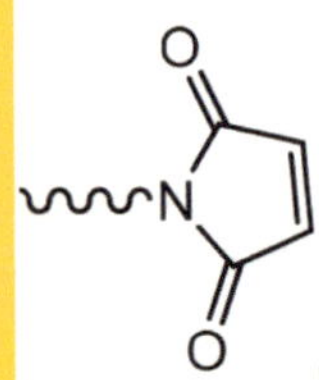		Maleimide/thiol [18, 19,30]
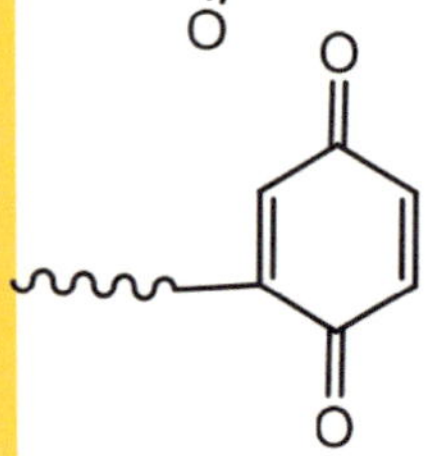		Benzoquinone/cyclopentadiene [17]
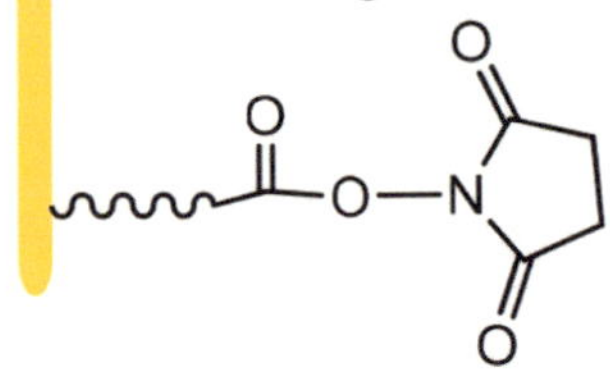		N-Hydroxysuccinimide-ester/amine [31]

Table 1. Different strategies to prepare SAMs of carbohydrate using indirect method.

cells/pmol of end groups, whereas generation three dendrimer detected 4170 cells/pmol of end groups [26].

Recently, the CuAAC along with amine/isothiocyanate click reactions were used by Grabosch and coworkers to prepare mannosylated SAMs on gold surfaces [27]. The surface was then used for selective recognition of GFP-transformed *E. coli* bacterial with a minimum of nonspecific interactions because of the prepared biorepulsive linker. There are many other pairs of functional groups [17–19, 28–31], which can react together for the successful preparation of carbohydrates SAM using indirect method. Some commonly used functional groups are listed in **Table 1**.

3.2. Direct method

The indirect method of forming SAM of carbohydrates has some advantages, but the direct method has also been widely used within last two decades. In this method, the carbohydrate (or glycan) of interest is directly functionalized with an organosulfur molecule, and the

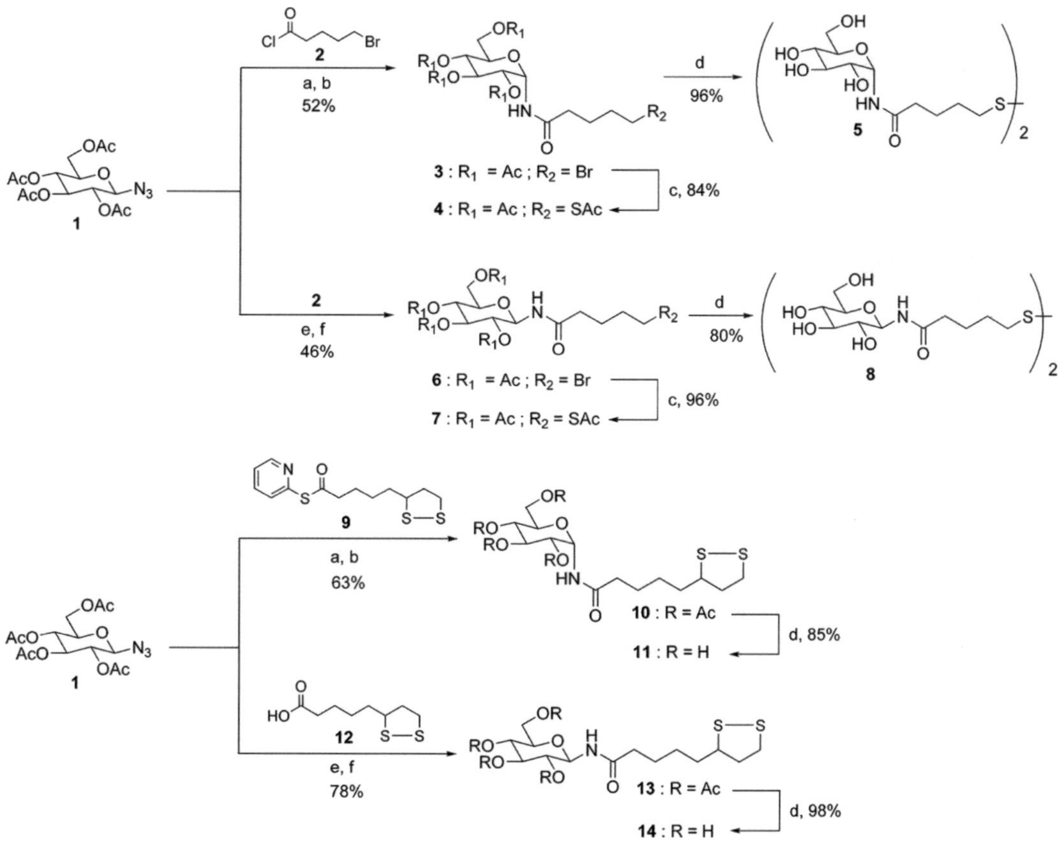

Figure 5. Scheme for synthesis of glucopyranosylamide conjugates. (a) PPh_3, 1,2-dichloroethane, 95°C, 16 h. (b) coupling partner, 24 h. (c) KSAc, DMF, RT, 20 h. (d) NaOMe, MeOH, RT. (e) PMe_3, DIEA, 1,2-dichloroethane, RT, 30 min. (f) coupling partner, RT, 24 h. Reprinted with permission from Ref. [34], Copyright 2007, American Chemical Society.

synthetic product is characterized and confirmed before forming SAMs. In the next step, freshly cleaned gold surface is incubated on ethanolic solution of the functionalized molecule for 1–24 h at ambient conditions to prepare the SAM. This method avoids multistep surface reactions. In 1996, Fritz and coworkers reported SAMs of a hexasaccharide molecule functionalized with alkanethiols on gold surfaces [32]. The conditions for high-density SAM formation were explored by performing experiments with or without protecting the hydroxyl groups with acetyl groups and determining whether unprotecting it after or before the immobilization gave the optimal condition. The optimal condition was found to be adsorption of the unprotected molecules from solution. The Russell laboratory later synthesized mannose-terminated alkanethiols to prepare SAMs and selectively capture Con A, and the expected selective capture of Con A was determined using reflection absorption infrared spectroscopy (RAIRS) and surface plasmon resonance (SPR); L-fucose-specific lectin *Tetragonolobus purpureas* was not captured [33]. The detailed synthetic strategies for the preparation of carbohydrate functionalized organosulfur SAM is outside the scope of this chapter. A representative scheme for preparation of the SAM of glucopyranosylamide is shown in **Figure 5**, which shows strategies for introducing different disulfide linkers to the acetyl protected glucose units having azide groups at the anomeric centers [34].

4. Types of terminal carbohydrates

It is possible to immobilize or form microarrays of a wide variety of carbohydrates on gold surfaces using SAM techniques. Different types of monosaccharides [17, 21, 27, 35–46], disaccharides [42, 43, 47–50], oligosaccharides [45, 46, 49–54], polysaccharides [14], and dendrimers [4, 5, 26] can be immobilized as a terminal functional group of the SAM depending on the nature of the study. **Table 2** lists some of the carbohydrates used as terminal functional groups of the SAM and their applications. The detailed applications of the carbohydrates will be discussed later.

Sacc.	Terminal carbohydrates	Applications
Monosaccharides	Glucose (α-D-Glc) 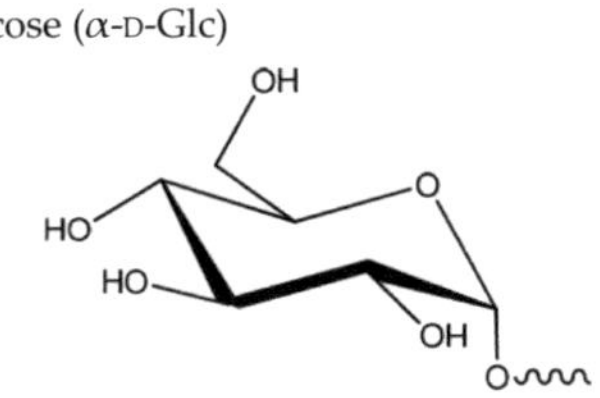	-to study the interaction with lectins [17, 55]
	Mannose (α-D-Man) 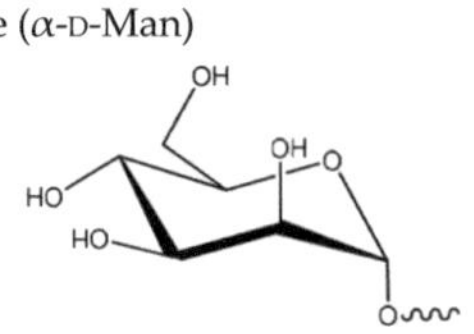	-to specifically detect *E. coli* bacteria [27, 56]-to load anti-HIV prodrug candidates on gold nanoparticles using "thiol-for-thiol" ligand place exchange reactions [57]-to study the interactions with lectin Con A to understand more complex carbohydrate-proteins [58, 59]

http://dx.doi.org/10.5772/66194

(cont.)

Sacc.	Terminal carbohydrates	Applications
	Galactose (α-D-Gal) 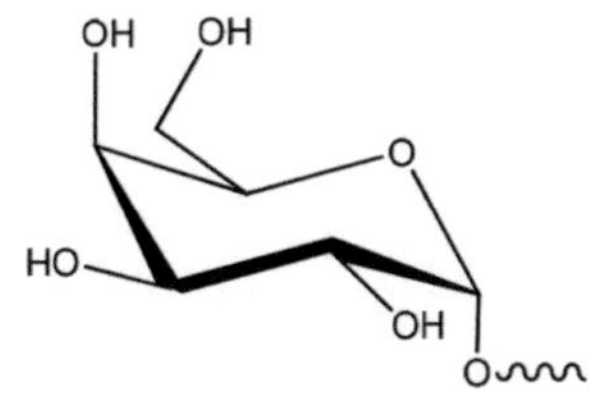	-to study the specific interactions with different types of lectins, for example, lectin from *E. cristagalli* [17], Jacalin [58], peanut agglutinin- to use as a proteins resist surface [60]-sensitive and facile detection of deadly plant protein, ricin [47]-to use as an anti-fouling surface for marine fouling organisms [35, 48]
	N-acetylglucosamine (GlcNAc) 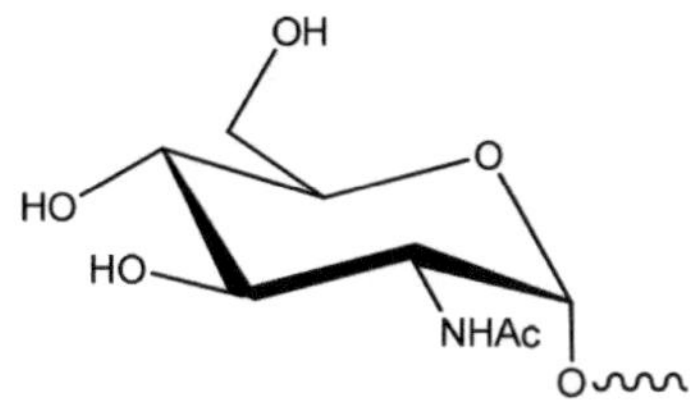	-wheat germ agglutinin binding study [36]- to study the enzymatic modification of the immobilized carbohydrates[37]
	N-acetylneuraminic acid (sialic acid) 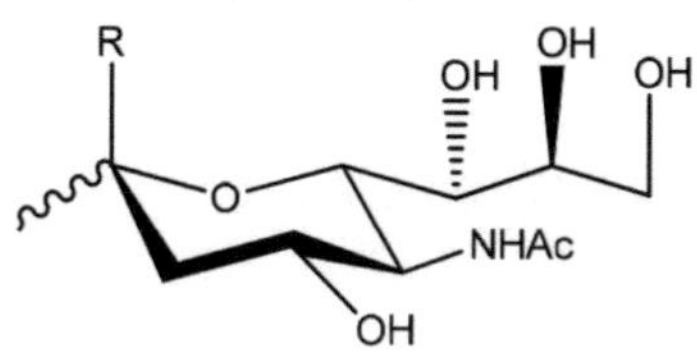 $R = COO^-$ or OH	-to detect Alzheimer's disease linked protein amyloid-beta (Aβ) at low concentration [21, 38]- to optically detect virus based on plasmonic properties of the sialic acid-linked gold nanoparticles assembled on the surface of viruses [39]
Disaccharides	Thomsen–Friedenreich antigen (TF_{ag})β-D-Gal-(1–4)-D-GalNAc 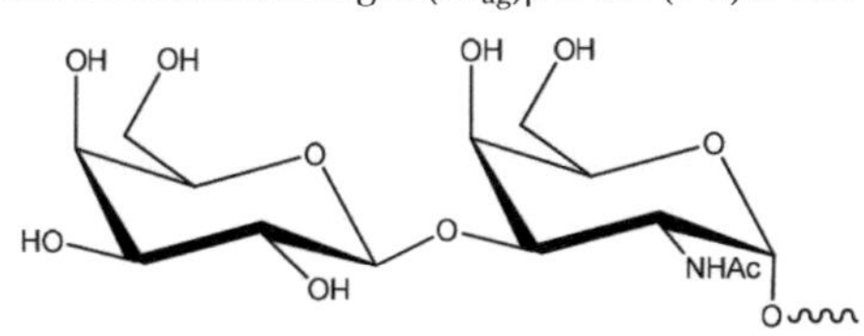	-to study binding properties and immune response of this tumor-associated carbohydrate antigen [40] and in antitumor therapeutics [41]
	Maltose 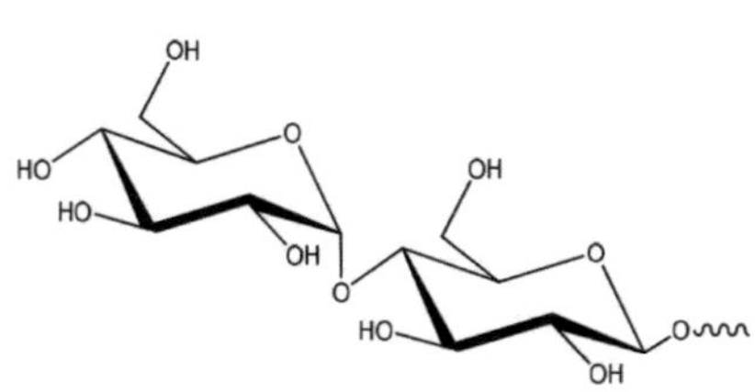	-to detect nanomolar concentration of Con A with high signal-to-noise ratio [42, 43]-to prepare antifouling surface against proteins and common marine fouling species [48]
	Lactose 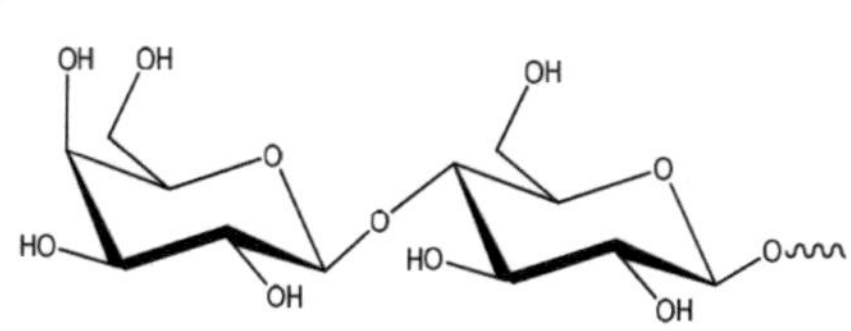	-to study the interaction with lectin *Erythrinacristagalli*-sensitive and facile detection of deadly plant proteinous toxin, ricin [47]-nM-level detection of galectins, a biomarker for cancer and other serious diseases [44, 45]

(cont.)

Sacc.	Terminal carbohydrates	Applications
Oligosaccharides	Globotriose (α-D-Gal-(1-4)-β-D-Gal-(1-4)-β-D-Glc) 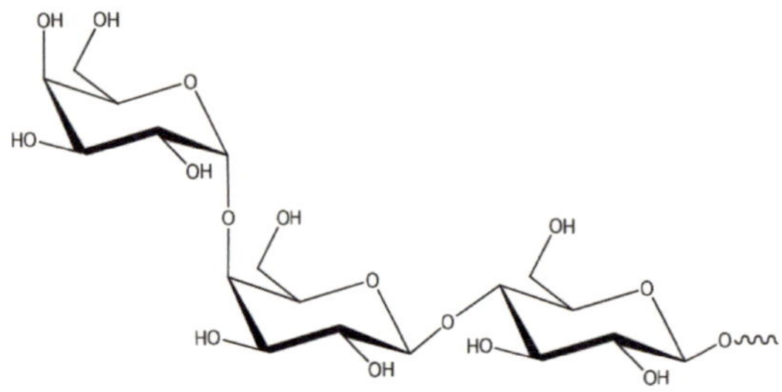	- to probe Shiga-like toxins[46]- as potential therapeutics for Shiga toxins when used with gold nanoparticles [50]-to discover the peptides that can inhibit Shiga toxin [49]
	Lewis x (le^{x}) 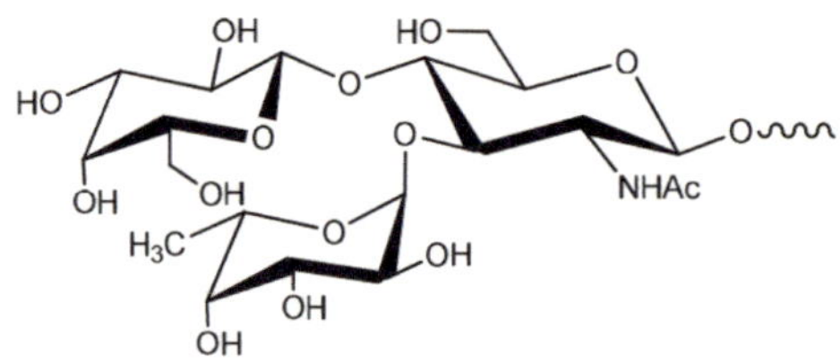	-to study the carbohydrate-carbohydrate interactions on gold nanoparticleSelf-recognition in aqueous solution was demonstrated by mimicking glycosphing olipid clusters [51, 52]
	Chitohexaose 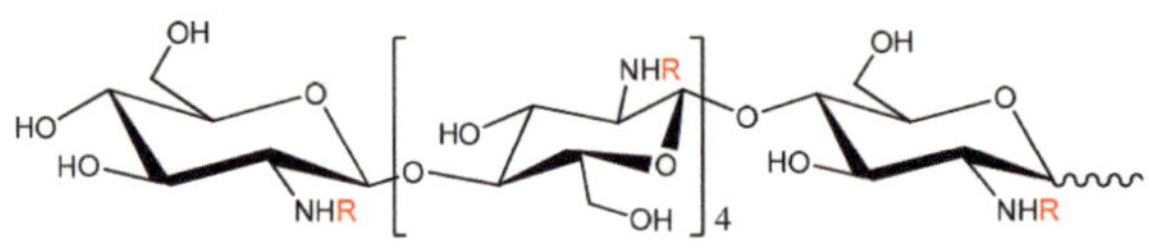 GlcNAc6: R = Ac, Glc6: R= H	-for immobilization of cells and use the surface for cell culture applications [53, 54]
	β-Cyclodextrin 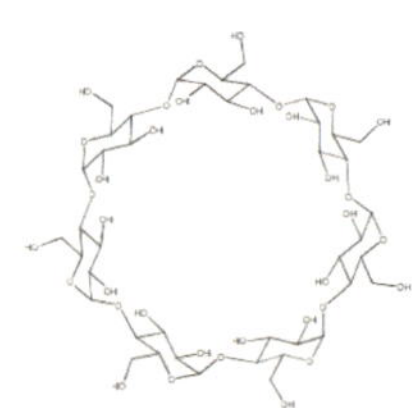	-to enhance the loading capacity of anticancer drug methotrexate on gold nanoparticles[45]
Dendrimers	Generation 2 dendrimer with terminal carbohydrates (**Figure 4**) Generation 3 dendrimer with terminal mannose group	-to determine association constants and kinetic rate constants with corresponding specific lectins [4, 5]-to enhance bacteria sensor [26]

Table 2. Commonly used carbohydrate molecules to prepare SAM on gold surface and their applications.

5. Role of linkers and mixed SAMs

The spacer connecting the sulfur atom to the terminal carbohydrates has an important role in the arrangement and application of SAMs. The most commonly used linkers are repeating units of a methylene group (-CH_2), ethylene glycol (-OCH_2CH_2) and the combination of both [29]. However, linkers can also be made from aromatic compounds [55], dendrimers [4, 5], and peptides [56]. Connecting the different types of linkers previously present on the thiols or carbohydrates may introduce complex structures to the linker; for example, triazoles from alkyne/azides reaction, thiourea-bridge from isothiocyanide/amine reaction and amides from N-hydroxyester/amine reaction. The main goals of the linkers are to provide strong support to the terminal group with a proper orientation, keep the terminal group far from the substrate, and resist non-specific interactions with proteins.

It has been reported that due to kinetic and thermodynamic reasons, longer chain linkers are relatively ordered and robust [11]. The length of the linker is also important for the arrangement of the SAM, which can significantly change the orientation of SAM as shown by Yatawara and coworkers [30]. Alkanethiol linkers of terminal glucose having 11 and 16 carbons chain have similar orientations, but are totally different compared to a cysteine containing linker. A study comparing the effect of thioctic acid amide and alkanethiol linkers on the interactions of terminal mannoses with specifically binding lectin Con A, non-specific lectins and with the highly adsorbent "sticky" proteins, fibrinogen and cytochrome c was performed [57]. The results show that the thioctic acid amide-based linker was better at resisting the nonspecific interactions while specifically binding terminal mannose to its corresponding lectin. It was claimed that this linker can resist adsorption of fibrinogen and cytochrome c better than ethylene oxide-based SAMs.

Even though the longer chain linkers are preferred due to various advantages, monolayers of oligosaccharides (e.g., hyaluronan, chitohexaose, and chitosan hexamer) have been successfully immobilized on gold by modifying the reducing ends of the carbohydrates with thiosemicarbazide (TSC) (**Figure 6**) [53, 58, 59]. The immobilized oligosaccharides SAMs were then used for specifically capturing different types of cell to be used in cell culture applications [53, 59, 60].

Mixed SAMs of carbohydrates are generally formed by two different constituents of thiolated molecules, one having terminal carbohydrates and other lacking carbohydrate molecules. The

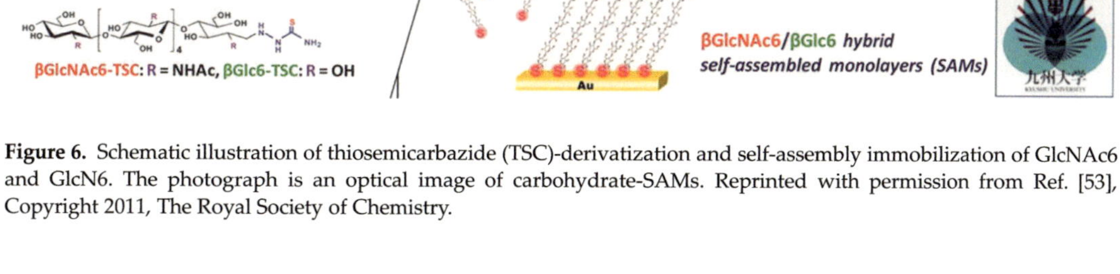

Figure 6. Schematic illustration of thiosemicarbazide (TSC)-derivatization and self-assembly immobilization of GlcNAc6 and GlcN6. The photograph is an optical image of carbohydrate-SAMs. Reprinted with permission from Ref. [53], Copyright 2011, The Royal Society of Chemistry.

most common way of making mixed SAMs of carbohydrates on gold surfaces is by co-adsorption of the two components from solution [61]. Other methods to prepare mixed SAMs are by adsorption of asymmetric disulfides on gold surfaces [18] and by ligand exchange reaction [62]. In the ligand exchange reaction, new thiolated molecules are introduced on the surface of the already formed SAMs by a thiol-for-thiol mechanism [62]. However, newly introduced molecules may not yield homogeneously mixed SAMs. The relative ratio of the component of the mixed SAM on the surface depends on the mole ratio of the components in the solution. However, increasing mole ratio of one component in solution does not necessarily increase its ratio on surface in a directly proportional manner [11].

A main goal of making a mixed SAM is to minimize non-specific interactions and create a biorepulsive background, as the hydrophobic chain of the linker might not be able to resist non-specific interactions. Mixed SAMs are also prepared to control the density of the terminal carbohydrates, as it has been found that crowding of receptors on the substrate surface is not an optimal condition for the binding of proteins or other analytes [63]. The crowded receptors may interact with each other or hinder the binding of the approaching analyte to the nearby receptors.

Mixed SAMs having two different types of terminal carbohydrates can also be created for a dual function. Aykaç and coworkers prepared SAMs of two different carbohydrates, lactose and β-cyclodextrin, on gold nanoparticles to selectively detect human galectin-3 through lactose whereas at the same time loading anticancer drug methotrexate on β-cyclodextrin [45]. This synergistic effect of two different terminal carbohydrates was found to be very effective for site-specific delivery of anticancer drug than when they are used individually.

6. Types of head-group

SAMs of carbohydrates on gold surfaces are extensively prepared based on thiol or disulfide head-groups. Thiols have higher solubility and normally form a well-ordered surface compared to disulfides [64]. However, they are susceptible to oxidation, forming sulfonates or disulfides and degrade over the time [11]. Disulfides, being the less soluble component, may precipitate out forming multilayer contamination if not prepared carefully [11]. In spite of this, disulfides are frequently used as head-groups for carbohydrate-terminated SAMs [65]. The carbohydrate-terminated SAMs formed by using dialkyl disulfide groups are found to be indistinguishable from those formed from the corresponding thiol and are believed to be formed by the cleavage of disulfide bond [66, 67]. However, such phenomenon is not very well studied or understood for the disulfides present in the cyclic form such as in the case of lipoic acid-based linkers. To reduce the problem of oxidation of thiols, they can be protected using different strategies and reduced in situ just before SAM formation. This can be done by keeping them as disulfides before and reducing them to corresponding thiols using dithiothreitol [68] or by first protecting the thiols using the S-trityl group followed by detritylation using trifluoroacetic acid and triethylsilane in dichloromethane [19].

http://dx.doi.org/10.5772/66194

7. Gold substrates for SAMs formation

Gold substrates are so far the most used and studied substrate for the formation of carbohydrate SAMs not only because they are capable of supporting stable SAMs due to Au-S bonding but also due to their conductivity, chemical and physical stability, and biocompatibility. SAMs of carbohydrates can be prepared on gold surfaces having different morphologies, such as planar (e.g., bulk or thin-films) to nanostructured surfaces (e.g., nanoparticles, nanostructured films, and nanoporous structures). Nanostructures of gold are intriguing to scientists because they can strongly scatter and absorb light due to large optical field enhancements [69] and have a high surface area-to-volume ratio [70] while still maintaining their other important bulk properties. Because of these properties, nanostructures of gold have application in diverse fields, including biomedicine (drug delivery), energy (hydrogen storage, solar cell, and battery), optics (sensors), and electronics (computer chips, information storage) [69, 71]. Carbohydrates SAM on nanoparticles can be prepared using direct and indirect methods similar to those explained before. However, they can also be prepared; at the same time, gold nanoparticles are prepared using reduction of gold salts by keeping the desired thiolated carbohydrates in the same solution mixture. The ligand exchange reaction is another way to introduce the SAM of desired carbohydrates to the already formed carbohydrate SAMs on nanoparticles surfaces [62]. SAMs prepared on nanoparticles may not be exactly same as the SAMs prepared on a planar surface due to the high radius of curvature of nanoparticles [11]. Since the carbohydrate immobilized nanoparticles are free to move around, they are used for studying the self-recognition of different carbohydrates [52], as possible inhibitors of lung cancer metastasis [72] and as an antitumor agent [41, 73]. SAMs of carbohydrates are also immobilized on other robust nanostructures of gold such as nanostructured gold film [74] and nanoporous gold (np-Au), which can be used as a biosensor transducer [75]. np-Au is a three-dimensional structure having pores (inter-ligament gaps) and ligaments with widths on the order of a few nanometers to a few hundreds of nanometers [70, 76]. np-Au was also used as a solid support for synthesizing disaccharides and trisaccharides starting from simple monosaccharide-terminated SAMs [77, 78].

8. Characterization techniques

There are wide varieties of methods to characterize the successful formation of SAM having terminated carbohydrates on gold surfaces and to study their interaction with other biomolecules. However, there is no single technique that alone can characterize the carbohydrates SAMs and their interactions completely, so different techniques should be used to support the result obtained from one method. Based on the purpose of the study, some of the most frequently used techniques are now discussed.

8.1. Surface wettability

The wettability of the surface before and after the modification by carbohydrate-terminated SAMs can be determined using contact angle goniometry by measuring the contact angle between water droplet and the surface [35]. The contact angle can be calculated by first taking images of the droplet of water on the surface and using software to fit different models. If contact angle is greater than 90°, the surface is considered hydrophobic; and if smaller than 90°, the surface is considered hydrophilic [79]. Unprotected carbohydrate-terminated surfaces typically create low contact angles owing to their hydrophilic nature. The static contact angle determined using a sessile droplet is the common method to check the surface wettability. However, due to the deviation from an ideal nature of the surface, there is always a contact angle hysteresis ranging from advancing contact angle to the receding contact angle [80]. The Liedberg group compared the wettability of the surface created with methylated and nonmethylated galactose-terminated SAMs on gold surfaces [81]. It was found that nonmethylated galactose surfaces had contact angle <10° demonstrating its hydrophilic properties, whereas methylated galactose surfaces had contact angle >70° demonstrating relative hydrophobicity. Dietrich and coworkers used contact angle goniometry to measure the contact angle of dimannoside-terminated SAMs on gold surfaces, which was found to be 36° ± 2° [79]. The reported contact angle is high compared to a pure hydroxyl-terminated surface, attributed to the exposed hydrophobic aliphatic linker. Fyrner and coworkers measured the advancing contact angle of oligo(lactose)-based thiol SAMs on gold, and it was found to be <10° [82]. This demonstrates a very hydrophilic surface as expected because of the highly hydroxylated oligosaccharides moieties.

8.2. Thickness and roughness

Ellipsometry is a powerful optical technique used for measuring the thickness and roughness of the carbohydrate-terminated SAM surface. This technique is based on changes in the polarization of incident radiation upon interacting with the surface of interest. The Konradsson group utilized this technique to study the increase in thickness of the oligo(lactose)-based SAM by introducing one, two, and three lactose units [82]. As expected, the thickness of the immobilized SAMs for one lactose unit 15.7 ± 0.2 is increased to 20.9 ± 0.2 for two units and to 29.8 ± 0.2 for three units. Another study using ellipsometry showed that $(CH_2)_{15}$ alkanethiol linker gave a height of 22 Å, which the addition of globotriose increased to 32 Å and on insertion of triethylene glycol between globotriose and the alkanethiol linker increased to 40 Å [68]. The results also showed that the thickness of the mixed monolayer increases nonlinearly with an increase in the ratio of the globotriose in solution, as the composition of globotriose in the incubation solution differs from that immobilized on the gold surface.

Atomic force microscopy (AFM) is another important tool, which can be used to directly image the sample to see the roughness on the surface with the resolution on the order of a fraction of a nanometer. It consists of cantilever having sharp tip made of silicon or silicon nitride that gets deflected upon interaction with the surface. Depending on the nature of study, deflecting force or force needed to keep the cantilever at a constant height can be measured using mainly two different modes, contact mode and non-contact mode (tapping mode). The Penades group prepared SAMs of disaccharide maltose on flat gold (111) surface as a neoglycoconjugate to mimic the glycolipids of cell membranes and characterized them using AFM [83]. They have found that despite the bulkiness of the carbohydrate groups, the SAMs were found to be well-ordered and densely packed comparable to SAMs of alkanethiols. A

http://dx.doi.org/10.5772/66194

similar conclusion was reached when α- or β-linked glucopyranosylamide derivatives SAMs were prepared on gold and studied using the contact mode of AFM [34]. The study has also found that acetate analogues of the same compound form multilayered films instead of forming SAMs under the similar deposition conditions. AFM was also used for monitoring the interactions of mannose-terminated SAMs to Con A by observing changes in roughness on the surface before and after the immobilization of Con A [24]. Increase in roughness and the average *z*-dimension value from ≈2.5 to ≈4.0 nm was observed after Con A immobilization. Relatively smaller height of Con A compared to the one obtained from X-ray crystallography (≈6.3–8.9 nm) was attributed to the tip-induced changes when contact mode was used. **Figure 7** shows work by Chikae and coworkers where AFM was used for monitoring changes in roughness on the surface after each modification step [21]. **Figure 7A** shows the AFM image of clean flat gold substrate, **7B** is the image after immobilization of sialic acid SAM, and **7C** and **7D** represent the images after capturing of Aβ(1–40) and Aβ(1–42), respectively after incubation of 20 μM Aβ peptides at room temperature for 180 min.

In a slightly different direction, AFM was also used for controlling the spacing of already immobilized carbohydrate SAMs by increasing the imaging force above the displacement threshold [84], which was then used to monitor the binding affinity of viral envelope glycoprotein gp120 to SAM of galactosylceramide prepared at controlled edge-to-edge gaps. The protein shows better immobilization when edge-to-edge separation of SAMs falls between 1.3 and 9.4 nm, with a 4.8 nm gap giving the optimal binding.

8.3. Chemical composition

X-ray photoelectron spectroscopy (XPS) is a powerful surface characterization technique, which provides useful information about elemental and chemical composition of carbohydrate SAMs. Cheng and coworkers utilized the XPS technique to determine the carbon, sulfur, and oxygen composition of diluting SAM, linker, and functionalized mannose [85]. In another study, high-resolution XPS analysis of C1s spectrum was performed by Dhayal and Ratner to identify the O-C-O functionality and separate/quantify the relative coverage of carbohydrate molecules in mixed SAMs [86]. They also found that the ratio of thiolated molecules in the solution is directly dependent on types of SAMs formed on the surface. High-resolution XPS was also used to track the nitrogen atom after the alkyne/azide click reaction [87]. The N1s spectra clearly does not show any peak before the click reaction, but after the click reaction, the N1s spectra shows a peak at 402 eV supporting the formation of triazole rings.

Infrared (IR) spectroscopy provides the information regarding the chemical composition present on the SAM surface based on frequencies and intensities of the molecular vibration. A simple carbohydrate molecule show characteristic bands in two different regions; the first is a broad band at around 3300–3500 cm^{-1} due to stretching of OH groups and the second is several modes between 1200 and 1000 cm^{-1} due to hemiacetal of carbohydrates (O-C-O) [13]. Increasing broadness of OH band and shifting to lower frequency indicate higher intermolecular and intramolecular hydrogen bonding, creating well-ordered densely packed monolayer formation [33]. Several other special molecules might be introduced through the linker when preparing the carbohydrate SAM that can be easily seen through the change in

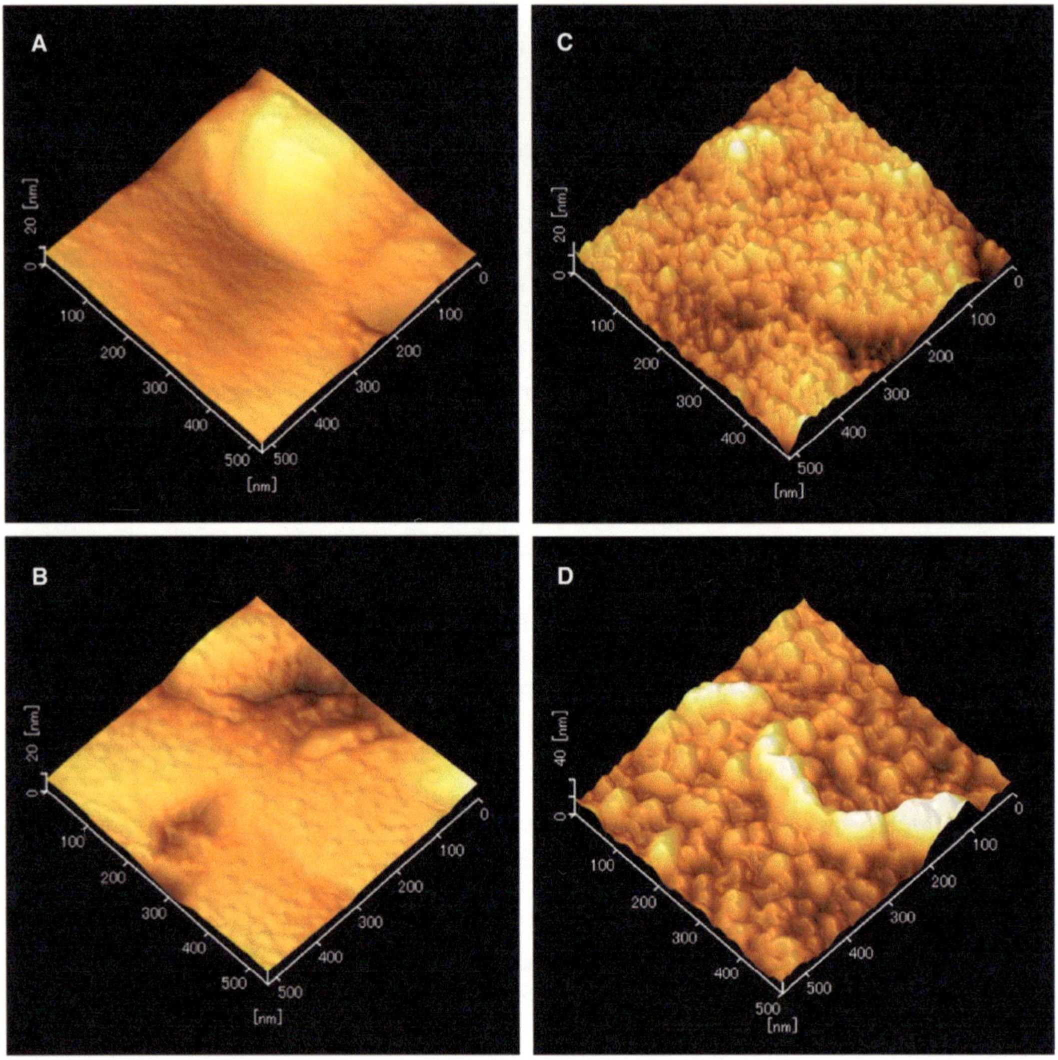

Figure 7. AFM images of the bare gold substrate (A); after cycloaddition of the sialic acid (B); the attachment of 20 μM Aβ peptides after incubation at RT for 180 min; Aβ(1–40) (C) and Aβ(1–42) (D). Reprinted with permission from Ref. [21], Copyright 2008, Elsevier.

position and sudden appearance or disappearance of the bands. Amide group present in the linker of the carbohydrate SAM was found to show bands at ≈1650 and ≈1560 cm^{-1} for amide I and amide II, respectively [34, 88]. Similarly, if the hydroxyl groups are protected with an acetyl group, a strong band will be seen at 1765 cm^{-1} due to ester functionalities.

8.4. Binding affinity

Surface plasmon resonance (SPR) is a popular biosensing tool based on change in the refractive index at the metal-sample interface. It is used for studying the affinity-based interactions of the biomolecules such as protein-protein interactions, label-free immunoassay, enzyme-substrate interactions, DNA hybridization, and diagnosis of virus-induced diseases in real time [89]. This method has also been widely used for studying the interactions of carbohydrate SAMs with proteins. Schlick and Cloninger studied the inhibition property of glycodendrimer for the binding of Con A to the SAM presenting mannose [90]. The equilibrium dissociation constant (K_d) of Con A to the mannose functionalized surface was initially determined to be 78 nM. The smaller K_d value for this interaction was attributed to multivalent interactions on the surface. A multivalent glycodendrimer framework was used to inhibit the interaction of mannose-functionalized dithiol SAMs on gold surfaces to Con A, whose IC_{50} values were found to be in the range from 260 to 13 nM. In another study, SPR was used for one-step detection of galectins, a β-galactoside-binding lectin well-known as a biomarker for different cancer, by preparing mixed SAMs of thiolated lactoside and triethylene glycol (TEG) [44]. The sensor designed is very sensitive that it can detect ≈ 1 nM galectin-4 and -8, in spite of very weak interactions of galectins and lactoside with $K_d \approx 1.0 \times 10^{-3}$ –1.0×10^{-6} M because of suppression of nonspecific interactions by TEG.

Localized surface plasmon resonance (LSPR) is a relative new biosensing technique. Scientists are exploring this technique because of its simplicity and possibility to be miniaturized, decreasing the application cost [91]. This technique is very sensitive and supports label-free a real-time biosensing. Similar to SPR, LSPR-based biosensing also depends upon the change in RI with output data commonly represented by measuring wavelength or intensity shift. LSPR-based biosensing, however, also depends on shape, size, and composition of the material used as a transducer [92]. The important step of LSPR-based biosensing is the fabrication of plasmonic sensitive metal nanostructures. Wide varieties of nanostructures have been created using different techniques with the goal of finding simple preparation methods, highly sensitive structures for detecting and studying biomolecules and their interactions, chemically and physically robust structures, and those that can be easily regenerated. Even though nanostructures of many noble metals can generate the LSPR signal, nanostructures of gold and silver are frequently used for LSPR-based biosensing [74]. Silver-based nanostructures show better sensitivity with sharper peaks than gold-based nanostructures [93]. However, silver nanostructures are prone to oxidation causing change in plasmonic properties and also weakening thiol-metal binding [94], making gold nanostructures the metal of choice for LSPR-based biosensing. Bellapadrona and coworkers prepared LSPR sensitive gold island films by evaporation of gold on glass followed by annealing [95]. As-prepared structures were used to form mannose SAM, and interaction with Con A was monitored by change in peak intensity and wavelength. The binding kinetics of mannose to Con A were also determined whose k_{on} and k_{off} are 2.0×10^4 M^{-1} s^{-1} and 2.6×10^3 s^{-1}, respectively. The detection limit of Con A on mannose-terminated SAM was achieved down to <5 nM. Previously, our group has also prepared a robust and sensitive nanostructured gold film (NGF) using a simple electrochemical method [74]. By immobilizing mixed SAMs presenting mannose, we were able to show the real-time interaction of Con A to mannose-terminated SAMs.

LSPR was applied to study monolayers of colloidal Au nanoparticles supported on glass. These were modified by polymer brushes presenting multiple glucose residues, and LSPR was used to determine a binding constant of $5.0 \pm 0.2 \times 10^5$ M^{-1} noted as larger than that for

Con A binding to methyl α-D-glucopyranoside of 2.4±0.1 × 10^3 M^{-1} in solution due to multivalent binding effects [96]. The use of supported gold nanoparticles modified with a polymer brush presenting many mannose units was also applied to follow Con A binding, resulting in an apparent association constant of 7.4 ′ 0.1 × 10^6 M^{-1}, noted as much greater than that for Con A to methyl α-D-mannopyranoside in solution of 7.6±0.2 × 10^3 M^{-1}, with the difference attributed to multivalent binding [97]. Galactose presenting polymer brushes was also used to modify colloidal gold monolayers and their binding of the lectin RCA120 was followed by LSPR, and the binding of HepG2 cells which contain galactose receptors was followed by optical microscopy [98]. The interaction of wheat germ agglutinin (WGA) with a disulfide-modified telomer polymer on a colloidal gold monolayer was also followed by LSPR [99].

Cyclic voltammetry: Cyclic voltammetry (CV) is one of the commonly used electroanalytical techniques for the study of SAMs. In this technique, the potential is linearly scanned back and forth making a triangular waveform in appropriate electrolyte solution. SAMs of carbohydrates and their interactions with biomolecules are normally determined by monitoring a decrease in peak current of a voltammogram. Li and coworkers used mannose-terminated SAMs to capture *E. coli* bacteria and the change in the peak current before and after the immobilization of bacteria was monitored [100]. It was found that the peak current decreased exponentially reaching a plateau within 5 min due to immobilization of *E. coli,* which means that there is a less free space surrounding the electrode for the probe molecules to reach the surface of the electrode. However, cyclic voltammetry may not always be a suitable technique to study the interactions, as the peak current may reach a plateau beforehand as the well-ordered SAM on gold surfaces leaves no further possibility for the peak current to decrease, which also means that if no probe molecules can penetrate the SAM, they also cannot penetrate any other biomolecules immobilized on SAM surface, and no change in current response will be seen [21, 75].

Electrochemical impedance spectroscopy (EIS) is another very sensitive technique used for biosensing [101]. This technique measures the resistive and capacitive properties of materials by applying a small alternating sinusoidal potential, typically 2–10 mV around a fixed potential usually chosen as the formal potential of a redox probe. An impedance spectrum is obtained by varying the frequency over a wide range and is most commonly represented in the form of a Nyquist plot [102]. Interactions of terminal carbohydrates of SAM with other biomolecules are observed by measuring the charge transfer resistance of the electrode before and after the modification of the surface. The Heineman group used EIS to detect the mannose-specific bacteria *E. coli* ORN 178 on mannose-terminated SAM on a gold surface [103]. With the incubation of *E. coli* ORN 178 on the mannose surface, the charge transfer resistance (R_{ct}) increased drastically in the Nyquist plot because of the selective binding, whereas incubating with *E. coli* ORN208 did not show any significant change in the signal supporting non-specificity toward mannose. Monitoring the change in signal with change in concentration of *E. coli* ORN 178, it proved possible to determine the detection capacity of the system, found to be from 10^2 to 10^3 CFU/mL. In another work, EIS was used to study the resistance of different carbohydrate-derived lipoic acid derivatives for nonspecific binding [65]. The interaction of single chain antibody (scFv) with rabbit IgG was studied on a gold surface. It was found that human serum and HeLa cell interfered with the interactions shown by an increase in charge transfer resistance. However, when galactose presenting SAM and bovine serum albumin (BSA) was prepared along with the scFv, the resistance decreased compared to scFv alone, but it did not increase upon passing human serum and HeLa cells, showing the effectiveness of the galactose SAM to resist background interactions. When antibody was passed over the SAM surface, it showed an increase in resistance due to selective binding.

Quartz crystal microbalance (QCM) is one of the frequently used techniques for biosensing in which a piezoelectric crystal, most commonly a quartz-crystal coated with a gold electrode, is made to vibrate at a particular frequency [104]. Depending on the increase in mass by the immobilization of the analyte to bioreceptor on the resonator surface, the resonance frequency of quartz crystal decreases, which can be measured electrically and the amount of mass change can be determined [105]. QCM is widely used to study carbohydrate SAMs and their interactions with different biomolecules. Zhang and coworkers used click reaction to immobilize

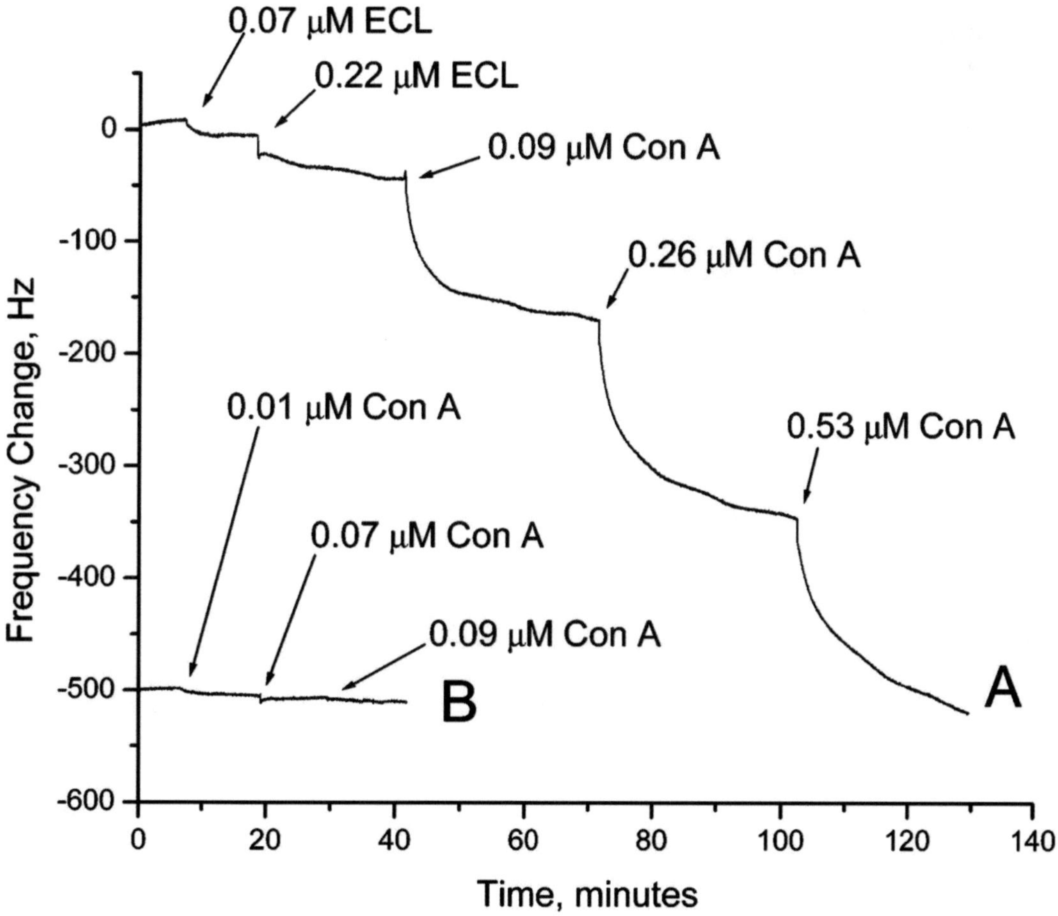

Figure 8. QCM analysis of *Erythrina cristagalli* lectins (ECLs) and Con A binding on (A) mannose SAM and (B) lactose SAM. ECL is specific to lactose but not to mannose, while Con A is specific to mannose but not to lactose. Reprinted with permission from Ref. [24], Copyright 2006, American Chemical Society.

azido sugars (mannose and lactose) on the alkyne-terminated SAM and studied their interactions with con A and *Erythrina cristagalli* lectins (ECLs) using QCM (**Figure 8**) [24]. It can be seen that ECL does not result in a change in frequency on the mannose modified resonator, whereas small concentration of Con A gives larger frequency change due to specific interaction with mannose. Similarly, Con A does not show any interaction with a lactose-functionalized resonator.

9. Applications

The carbohydrates present on the cells surface can act as a receptor for many pathogens to facilitate cell-cell adhesion through which humans can be infected, for example, mannose binds pathogenic bacteria *E. coli* and sialic acid binds influenza virus. Therefore, cell surface can be mimicked by preparing SAMs of carbohydrates to study and understand different types of binding and inhibition studies in vitro. SAMs of carbohydrates are also used for detecting disease biomarkers and carrying drugs.

9.1. Carbohydrate–lectin interactions

The diverse arrangement of carbohydrate in biological molecules makes their study challenging. However, the ubiquitous presence of 10 common monosaccharaides, namely, D-glucose (D-Glc), D-mannose (D-Man), D-galactose (D-Gal), N-acetylglucosamine (D-GlcNAc), N-acetylgalactosamine (D-GalNAc), D-glucuronic acid (D-GlcA), L-fucose (L-Fuc), N-acetylneuraminic acid (Neu5Ac), D-xylose (D-Xyl) and L-iduronic acid (L-IdoA), has made it easier to understand these structures and their functions, mainly by selecting lectins specific to these monosaccharides. Lectins are the proteins having an ability to bind specific types of carbohydrate [106] and hence to variety of glycoproteins, bacteria, and viruses through their carbohydrate-binding moieties [107]. Examples of commonly used lectins include Concanavalin A (Con A)-specific to mannose and glucose [108], peanut agglutinin (PNA) and jacalin-specific to galactosyl (β-1,3) N-acetylgalactosamine sugar sequence [109] and wheat germ agglutinin (WGA)-specific to N-acetylglucosamine.

Carbohydrate-lectin interactions can be studied in solution using techniques like isothermal titration calorimetry [110] or on gold surfaces using SPR, LSPR, EIS, and QCM [75, 104]. It has been found that the interactions between carbohydrate and lectin are stronger when performed on solid surfaces. The reason behind this is the favorable multivalency condition on the solid surface [111]. However, care should be taken when studying the interactions on solid surface as the defects on immobilized film can cause the analyte (protein) to immobilize directly on solid surface and can also precipitate the protein. The main goal of the carbohydrate-protein interactions study is to find the binding constant or to detect protein at low concentration. The change in response before and after the interactions of carbohydrate and lectin mostly in the form of optical, electrochemical, thermal or mass response is recorded. Then the change in response is recorded for wide range of concentrations creating a calibration plot, from which binding kinetics can be determined. The lower the value of K_d, the stronger is the binding between carbohydrate and lectin. K_d is the important information needed to guide the preparation of different inhibitors.

Since monosaccharides are easier to be derivatized to prepare SAMs, their interactions with their corresponding lectins have been well explored using different techniques. For example, K_d of mannose and Con A interactions are reported in the range of tens of nanomolar to few hundreds of nanomolar [112]. The variation on K_d arises due to the number of factors, including techniques used, type of substrate and SAM, preparation method, and functional activity of lectins. Loaiza and coworkers prepared screen-printed carbon electrodes modified with gold and functionalized with D-glucose and D-mannose SAMs [113]. The substrate was then used for detecting Con A using an electrochemical impedimetric technique finding detection limits of 0.099 and 0.078 pmol for D-glucose and D-mannose SAMs, respectively.

9.2. SAMs of carbohydrates for the detection of *E. coli*

Escherichia coli (E. coli) is a gram negative, rod-shaped bacteria naturally found in the intestines of humans and other animals. Many subset of *E. coli* are harmless; however, they are also the major cause of diarrheal disease among children in low-income countries [114]. The subset of *E. coli* uropathogenic *E. coli* (UPEC) is a leading cause of urinary tract infection (UTI) in humans [115]. According to the National Kidney Foundation, there are nearly 10 million doctor visits each year due to UTI, and one in five women will have it at least once in a lifetime. *E. coli* uses the FimH adhesin present at the tip of the type 1 pili to mediate adherence and invasion to urothelial cells through D-mannose groups present on the cell surface [116]. Invasion of *E. coli* into urothelial cells can not only lead to substantial medical cost for treatment but also damage different parts of the urinary system, including kidney creating potentially life-threatening complication if untreated. The culture and colony counting method is one of the preferred methods for the detection of bacteria, but it normally takes days to obtain the results. Research for developing simple, sensitive, and selective techniques, which can detect, with or without labeling, bacteria cells in shorter periods of time at cheaper price, are ongoing. Mannose microarrays prepared using SAM technique is a promising way to prepare such types of detector. Some of the previously reported work based on *E. coli* detection on mannose SAM surface is discussed below.

Understanding the microbial force of adhesion to the carbohydrate surface can help develop a new approach for detection and prevention of bacterial infection by blocking or decreasing the adhesion capacity. The forces of adhesion of UPEC to the mannose presenting SAM surface, representing the surface of epithelial cells, have been studied using optical tweezers by Whitesides group [117]. The group was successful to orient the bacteria end-on on mannose surface, from where they can be immediately detached and reattached onto mannose surface and the force required to detach from the surface was measured. In another study, SAMs of octadecanethiol on a gold electrode surface and polydiacetylene derivatives with or without terminal mannose were used to prepare bilayer similar to the biological membrane. Incubation of the prepared electrode in *E. coli* solution having 9×10^8 cells/mL for 5 min changes the initial blue color of the electrode to red, which can simply be observed through naked eyes. The change was further confirmed by resonance Raman spectroscopy,

UV-vis and EIS. Transmission electron microscope (TEM) was used to detect *E. coli* ORN 178 strain and distinguish it from non-mannose binding strain *E coli* ORN 208 on gold nanoparticles (AuNP) encapsulated by SAM of mannose. TEM micrographs of the area of pili clearly show selectively bound mannose encapsulated AuNP on pili of *E. coli* ORN 178 strain, but not on pili of the *E. coli* ORN 208 strain [118]. Recently, detection and comparison study of *E. coli* ORN 178 and *E. coli* ORN 208 was also performed using EIS technique where α-mannoside-terminated SAM was immobilized on a gold disk electrode [103]. The system is sensitive to detect bacteria in the range from 10^2 to 10^3 CFU/mL. To make label free and sensitive sensor for detecting bacteria, the synergistic combination of mannose SAM and Con A was utilized as a molecular recognition unit (**Figure 9**) [104]. Compared to the direct interaction of SAM of mannose to *E coli*, this synergistic combination showed significant improvement in attachment, sensitivity and specificity when monitored through the quartz crystal microbalance (QCM) transducer.

The adhesion of Con A to *E. coli* was through the multiple binding to lipopolysaccharides (O-antigen) exposed on cell wall of *E. coli*. LOD for mannose/Con A sensor was improved down to 7.5×10^2 cells/mL from 3.0×10^7 cells/mL for mannose-alone sensor. Similarly, four decades wider linear range ($7.5 \times 10^2 - 7.5 \times 10^7$ cells/mL) was found for the mannose/Con A-based sensor compared to mannose-alone QCM sensor [104]. In detecting *E. coli* on mannose-terminated SAMs, it is always preferred to have minimum nonspecific interaction for better selectivity and specificity of the detector. Grabosch et al. introduced a dual click chemistry strategy for creating a biorepulsive background with the exposed mannose terminal SAMs [27]. A polyethylene glycol linker having azide and amine terminal groups was used to support dual click reactions. Azides react with alkynes to form triazole ligation products and amines react with isothiocyanate to form thiourea bridges where the other ends of alkyne and isothiocyanate consist of thiol head group and terminal mannose, respectively. The SAMs of mannose prepared by this dual click strategy was found to be very effective in reducing the nonspecific interactions while specifically and selectively capturing green fluorescent protein (GFP)-tagged *E. coli* strain (pPKL1162) evident by epifluorescence micrographs with and without

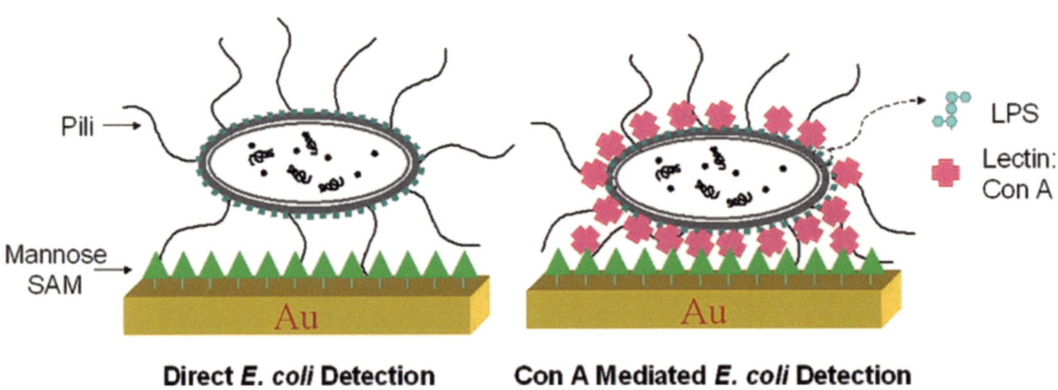

Figure 9. Schematic presentation of direct *E. coli* detection and Con A-mediated *E. coli* detection. Reprinted with permission from Ref. [104], Copyright 2007, American Chemical Society.

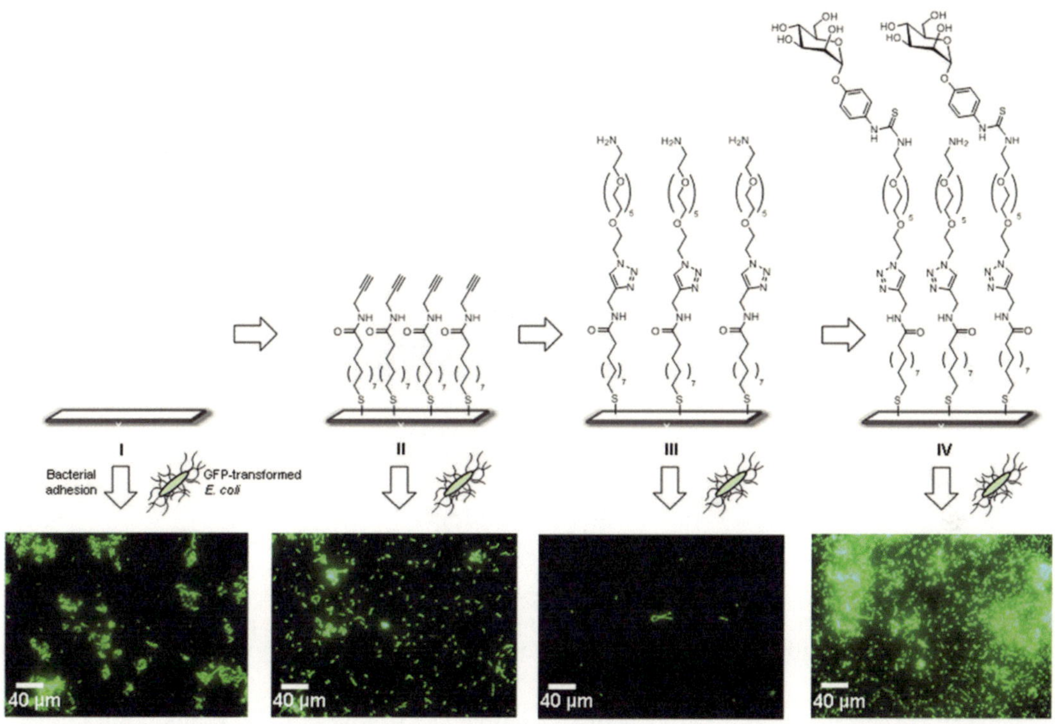

Figure 10. Adhesion of fluorescent bacteria to the different stages of the SAM during the 'dual click' approach. The GFP-transformed *E. coli* bacteria (pPKL1162) enable a fast, direct fluorescence readout to investigate bacterial adhesion on surfaces. The native gold surface (I) was used as reference in each of the other experiments. As can be seen in the epifluorescence micrographs, the (non-specific) adhesivity of the alkyne-terminated SAM II is comparable to the one of the native Au surface. Introduction of the OEG chain reduces the adhesion significantly, while the α-mannosyl-terminated SAM is effectively recognized by the *E. coli* leading to heavy adhesion. Reprinted with permission from Ref. [27], Copyright 2013, The Royal Society of Chemistry.

terminal mannose (**Figure 10**). The same strand of *E. coli* was also successfully detected on self-assembled dendritic monolayer (SADM) having disulfide cores using SPR and resonance-enhanced surface impedance (RESI) [26]. Generation one dendrimer having four terminal mannoses was found to have the binding efficiency of approximately 3600 cells·(pmol of Man)$^{-1}$ whereas generation three dendrimer having 16 terminal mannose have binding efficiency of 4200 cells·(pmol of Man)$^{-1}$. In this work, binding efficiency of generation three dendrimer having terminal mannoses and hydroxides groups to *E. coli* was also compared showing that mannose-terminated surface can improve the attachment of cells by 2.5-fold.

10. Summary

SAMs of carbohydrate can be prepared on gold surfaces to present multivalency and mimic the cell surface to study different physiologically significant interaction in vitro. In this chapter, we have presented the direct and indirect methods for forming SAM of carbohydrates on gold surfaces. Common strategies of preparing SAMs using indirect method are discussed and presented. We have also tabulated some of the commonly used carbohydrate terminal groups of the SAM, and their applications are presented. Different characterization techniques based on nature of study were also presented. Finally, the application of carbohydrates SAM for lectin and bacteria detection has been discussed.

Acknowledgements

The authors acknowledge recent support of their work in this area by University of Missouri–St. Louis and by the NIGMS awards R01-GM090254 and R01-GM111835.

Author details

Jay K. Bhattarai[1,2], Dharmendra Neupane[1,2], Vasilii Mikhaylov[1,2], Alexei V. Demchenko[1] and Keith J. Stine[1,2*]

*Address all correspondence to: kstine@umsl.edu

1 Department of Chemistry and Biochemistry, University of Missouri–St. Louis, Saint Louis, MO, USA

2 Center for Nanoscience, University of Missouri–St. Louis, Saint Louis, MO, USA

References

[1] Weymouth-Wilson, AC. The role of carbohydrates in biologically active natural products. Natural Product Reports. 1997;14(2):99–110.

[2] Liu, J, Willför, S, Xu, C. A review of bioactive plant polysaccharides: biological activities, functionalization, and biomedical applications. Bioactive Carbohydrates and Dietary Fibre. 2015;5(1):31–61.

[3] Kang, B, Opatz, T, Landfester, K, Wurm, FR. Carbohydrate nanocarriers in biomedical applications: functionalization and construction. Chemical Society Reviews. 2015;44(22):8301–8325.

[4] Fukuda, T, Onogi, S, Miura, Y. Dendritic sugar-microarrays by click chemistry. Thin Solid Films. 2009;518(2):880–888.

[5] Bogdan, N, Roy, R, Morin, M. Glycodendrimer coated gold nanoparticles for proteins detection based on surface energy transfer process. RSC Advances. 2012;2(3):985–991.

[6] Nicosia, C, Huskens, J. Reactive self-assembled monolayers: from surface functionalization to gradient formation. Materials Horizons. 2014;1(1):32–45.

[7] Ulman, A. Formation and structure of self-assembled monolayers. Chemical Reviews. 1996;96(4):1533–1554.

[8] Hakkinen, H. The gold-sulfur interface at the nanoscale. Nature Chemistry. 2012;4 (6):443–455.

[9] Laibinis, PE, Whitesides, GM, Allara, DL, Tao, YT, Parikh, AN, Nuzzo, RG. Comparison of the structures and wetting properties of self-assembled monolayers of n-alkanethiols on the coinage metal surfaces, copper, silver, and gold. Journal of the American Chemical Society. 1991;113(19):7152–7167.

[10] Mrksich, M, Whitesides, GM. Using self-assembled monolayers to understand the interactions of man-made surfaces with proteins and cells. Annual Review of Biophysics. 1996;25:55–78.

[11] Love, JC, Estroff, LA, Kriebel, JK, Nuzzo, RG, Whitesides, GM. Self-assembled monolayers of thiolates on metals as a form of nanotechnology. Chemical Reviews. 2005;105 (4):1103–1170.

[12] Evans, SD, Sharma, R, Ulman, A. Contact angle stability: reorganization of monolayer surfaces? Langmuir. 1991;7(1):156–161.

[13] Leone, G, Consumi, M, Lamponi, S, Magnani, A. Combination of static time of flight secondary ion mass spectrometry and infrared reflection–adsorption spectroscopy for the characterisation of a four steps built-up carbohydrate array. Applied Surface Science. 2012;258(17):6302–6315.

[14] Lofas, S. Dextran modified self-assembled monolayer surfaces for use in biointeraction analysis with surface plasmon resonance. Pure and Applied Chemistry. 1995;67(5):829–834.

[15] Liang, L, Astruc, D. The copper(I)-catalyzed alkyne-azide cycloaddition (CuAAC) "click" reaction and its applications: an overview. Coordination Chemistry Reviews. 2011;255 (23–24):2933–2945.

[16] Himo, F, Lovell, T, Hilgraf, R, Rostovtsev, VV, Noodleman, L, Sharpless, KB, Fokin, VV. Copper (I)-catalyzed synthesis of azoles: DFT study predicts unprecedented reactivity and intermediates. Journal of the American Chemical Society. 2005;127(1):210–216.

[17] Houseman, BT, Mrksich, M. Carbohydrate arrays for the evaluation of protein binding and enzymatic modification. Cell Chemical Biology. 2002;9(4):443–454.

[18] Houseman, BT, Gawalt, ES, Mrksich, M. Maleimide-functionalized self-assembled monolayers for the preparation of peptide and carbohydrate biochips. Langmuir. 2003;19 (5):1522–1531.

[19] Wehner, JW, Weissenborn, MJ, Hartmann, M, Gray, CJ, Sardzik, R, Eyers, CE, Flitsch, SL, et al. Dual purpose S-trityl-linkers for glycoarray fabrication on both polystyrene and gold. Organic & Biomolecular Chemistry. 2012;10(44):8919–8926.

[20] Bouchet-Spinelli, A, Reuillard, B, Coche-Guerente, L, Armand, S, Labbe, P, Fort, S. Oligosaccharide biosensor for direct monitoring of enzymatic activities using QCM-D. Biosensors and Bioelectronics. 2013;49:290–296.

[21] Chikae, M, Fukuda, T, Kerman, K, Idegami, K, Miura, Y, Tamiya, E. Amyloid-β detection with saccharide immobilized gold nanoparticle on carbon electrode. Bioelectrochemistry. 2008;74(1):118–123.

[22] Chelmowski, R, Kafer, D, Koster, SD, Klasen, T, Winkler, T, Terfort, A, Metzler-Nolte, N, et al. Postformation modification of SAMs: using click chemistry to functionalize organic surfaces. Langmuir. 2009;25(19):11480–11485.

[23] Sun, XL, Stabler, CL, Cazalis, CS, Chaikof, EL. Carbohydrate and protein immobilization onto solid surfaces by sequential Diels-Alder and azide-alkyne cycloadditions. Bioconjugation Chemistry. 2006;17(1):52–57.

[24] Zhang, Y, Luo, S, Tang, Y, Yu, L, Hou, KY, Cheng, JP, Zeng, X, et al. Carbohydrate-protein interactions by "clicked" carbohydrate self-assembled monolayers. Analytical Chemistry. 2006;78(6):2001–2008.

[25] Matsumoto, E, Yamauchi, T, Fukuda, T, Miura, Y. Sugar microarray via click chemistry: molecular recognition with lectins and amyloid β (1–42). Science and Technology of Advanced Materials. 2009;10(3):034605.

[26] Oberg, K, Ropponen, J, Kelly, J, Lowenhielm, P, Berglin, M, Malkoch, M. Templating gold surfaces with function: a self-assembled dendritic monolayer methodology based on monodisperse polyester scaffolds. Langmuir. 2013;29(1):456–465.

[27] Grabosch, C, Kind, M, Gies, Y, Schweighofer, F, Terfort, A, Lindhorst, TK. A 'dual click' strategy for the fabrication of bioselective, glycosylated self-assembled monolayers as glycocalyx models. Organic and Biomolecular Chemistry. 2013;11(24):4006–4015.

[28] Zhi, Z, Powell, A, Turnbull, J. Fabrication of carbohydrate microarrays on gold surfaces: direct attachment of nonderivatized oligosaccharides to hydrazide-modified self-assembled monolayers. Analytical Chemistry. 2006;78(14):4786–4793.

[29] Cheng, F, Ratner, DM. Glycosylated self-assembled monolayers for arrays and surface analysis. Carbohydrate Microarrays: Methods and Protocols. 2012:87–101.

[30] Yatawara, AK, Tiruchinapally, G, Bordenyuk, AN, Andreana, PR, Benderskii, AV. Carbohydrate surface attachment characterized by sum frequency generation spectroscopy. Langmuir. 2009;25(4):1901–1904.

[31] Zhi, ZL, Laurent, N, Powell, AK, Karamanska, R, Fais, M, Voglmeir, J, Wright, A, et al. A versatile gold surface approach for fabrication and interrogation of glycoarrays. ChemBiochem. 2008;9(10):1568–1575.

[32] Fritz, MC, Hähner, G, Spencer, ND, Bürli, R, Vasella, A. Self-assembled hexasaccharides: surface characterization of thiol-terminated sugars adsorbed on a gold surface. Langmuir. 1996;12(25):6074–6082.

[33] Revell, DJ, Knight, JR, Blyth, DJ, Haines, AH, Russell, DA. Self-assembled carbohydrate monolayers: formation and surface selective molecular recognition. Langmuir. 1998;14(16):4517–4524.

[34] Kadalbajoo, M, Park, J, Opdahl, A, Suda, H, Kitchens, CA, Garno, JC, Batteas, JD, et al. Synthesis and structural characterization of glucopyranosylamide films on gold. Langmuir. 2007;23(2):700–707.

[35] Ederth, T, Ekblad, T, Pettitt, ME, Conlan, SL, Du, C-X, Callow, ME, Callow, JA, et al. Resistance of galactoside-terminated alkanethiol self-assembled monolayers to marine fouling organisms. ACS Applied Materials & Interfaces. 2011;3(10):3890–3901.

[36] Lienemann, M, Paananen, A, Boer, H, de la Fuente, JM, Garcia, I, Penades, S, Koivula, A. Characterization of the wheat germ agglutinin binding to self–assembled monolayers of neoglycoconjugates by AFM and SPR. Glycobiology. 2009;19(6):633–643.

[37] Su, J, Mrksich, M. Using mass spectrometry to characterize self-assembled monolayers presenting peptides, proteins, and carbohydrates. Angewandte Chemie International Edition. 2002;41(24):4715–4718.

[38] Beier, HT, Cowan, CB, Chou, I-H, Pallikal, J, Henry, JE, Benford, ME, Jackson, JB, et al. Application of surface-enhanced Raman spectroscopy for detection of beta amyloid using nanoshells. Plasmonics. 2007;2(2):55–64.

[39] Niikura, K, Nagakawa, K, Ohtake, N, Suzuki, T, Matsuo, Y, Sawa, H, Ijiro, K. Gold nanoparticle arrangement on viral particles through carbohydrate recognition: a non-cross-linking approach to optical virus detection. Bioconjugate Chemistry. 2009;20 (10):1848–1852.

[40] Sundgren, A, Barchi, JJ. Varied presentation of the Thomsen–Friedenreich disaccharide tumor-associated carbohydrate antigen on gold nanoparticles. Carbohydrate Research. 2008;343(10):1594–1604.

[41] Biswas, S, Medina, SH, Barchi, JJ, Jr. Synthesis and cell-selective antitumor properties of amino acid conjugated tumor-associated carbohydrate antigen-coated gold nanoparticles. Carbohydrate Research. 2015;405:93–101.

[42] Sato, Y, Yoshioka, K, Tanaka, M, Murakami, T, Ishida, MN, Niwa, O. Recognition of lectin with a high signal to noise ratio: carbohydrate-tri(ethylene glycol)-alkanethiol co-adsorbed monolayer. Chemical Communications (Cambridge, England). 2008(40):4909–4911.

[43] Sato, Y, Murakami, T, Yoshioka, K, Niwa, O. 12-Mercaptododecyl β-maltoside-modified gold nanoparticles: specific ligands for concanavalin A having long flexible hydrocarbon chains. Analytical and Bioanalytical Chemistry. 2008;391(7):2527–2532.

[44] Yoshioka, K, Sato, Y, Murakami, T, Tanaka, M, Niwa, O. One-step detection of galectins on hybrid monolayer surface with protruding lactoside. Analytical Chemistry. 2010;82 (4):1175–1178.

[45] Aykac, A, Martos-Maldonado, MC, Casas-Solvas, JM, Quesada-Soriano, I, Garcia-Maroto, F, Garcia-Fuentes, L, Vargas-Berenguel, A. beta-Cyclodextrin-bearing gold glyconanoparticles for the development of site specific drug delivery systems. Langmuir. 2014;30(1):234–242.

[46] Chien, YY, Jan, MD, Adak, AK, Tzeng, HC, Lin, YP, Chen, YJ, Wang, KT, et al. Globotriose-functionalized gold nanoparticles as multivalent probes for Shiga-like toxin. ChemBioChem. 2008;9(7):1100–1109.

[47] Uzawa, H, Ohga, K, Shinozaki, Y, Ohsawa, I, Nagatsuka, T, Seto, Y, Nishida, Y. A novel sugar-probe biosensor for the deadly plant proteinous toxin, ricin. Biosensors and Bioelectronics. 2008;24(4):923–927.

[48] Fyrner, T, Lee, H-H, Mangone, A, Ekblad, T, Pettitt, ME, Callow, ME, Callow, JA, et al. Saccharide-functionalized alkanethiols for fouling-resistant self-assembled monolayers: synthesis, monolayer properties, and antifouling behavior. Langmuir. 2011;27(24):15034–15047.

[49] Miura, Y, Sasao, Y, Kamihira, M, Sakaki, A, Iijima, S, Kobayashi, K. Peptides binding to a Gb3 mimic selected from a phage library. Biochimica et Biophysica Acta (BBA)-General Subjects. 2004;1673(3):131–138.

[50] Kulkarni, AA, Fuller, C, Korman, H, Weiss, AA, Iyer, SS. Glycan encapsulated gold nanoparticles selectively inhibit Shiga Toxins 1 and 2. Bioconjugate Chemistry. 2010;21 (8):1486–1493.

[51] de La Fuente, JM, Barrientos, AG, Rojas, TC, Rojo, J, Canada, J, Fernandez, A, Penades, S. Gold glyconanoparticles as water-soluble polyvalent models to study carbohydrate interactions. Angewandte Chemie International Edition. 2001;40(12):2257–2261.

[52] Hernáiz, MJ, de la Fuente, JM, Barrientos, ÁG, Penadés, S. A model system mimicking glycosphingolipid clusters to quantify carbohydrate self-interactions by surface plasmon resonance. Angewandte Chemie. 2002;114(9):1624–1627.

[53] Yoshiike, Y, Kitaoka, T. Tailoring hybrid glyco-nanolayers composed of chitohexaose and cellohexaose for cell culture applications. Journal of Materials Chemistry. 2011;21 (30):11150–11158.

[54] Poosala, P, Kitaoka, T. Chitooligomer-immobilized biointerfaces with micropatterned geometries for unidirectional alignment of myoblast cells. Biomolecules. 2016;6(1):12.

[55] Seo, JH, Adachi, K, Lee, BK, Kang, DG, Kim, YK, Kim, KR, Lee, HY, et al. Facile and rapid direct gold surface immobilization with controlled orientation for carbohydrates. Bioconjugation Chemistry. 2007;18(6):2197–2201.

[56] Kaplan, JM, Shang, J, Gobbo, P, Antonello, S, Armelao, L, Chatare, V, Ratner, DM, et al. Conformationally constrained functional peptide monolayers for the controlled display of bioactive carbohydrate ligands. Langmuir. 2013;29(26):8187–8192.

[57] Karamanska, R, Mukhopadhyay, B, Russell, DA, Field, RA. Thioctic acid amides: convenient tethers for achieving low nonspecific protein binding to carbohydrates presented on gold surfaces. Chemical Communications. 2005(26):3334–3336.

[58] Yokota, S, Kitaoka, T, Sugiyama, J, Wariishi, H. Cellulose I nanolayers designed by self-assembly of its thiosemicarbazone on a gold substrate. Advanced Materials. 2007;19 (20):3368–3370.

[59] Tanaka, N, Yoshiike, Y, Yoshiyama, C, Kitaoka, T. Self-assembly immobilization of hyaluronan thiosemicarbazone on a gold surface for cell culture applications. Carbohydrate Polymers. 2010;82(1):100–105.

[60] Yoshiike, Y, Yokota, S, Tanaka, N, Kitaoka, T, Wariishi, H. Preparation and cell culture behavior of self-assembled monolayers composed of chitohexaose and chitosan hexamer. Carbohydrate Polymers. 2010;82(1):21–27.

[61] Horan, N, Yan, L, Isobe, H, Whitesides, GM, Kahne, D. Nonstatistical binding of a protein to clustered carbohydrates. Proceedings of the National Academy of Sciences of the United States of America. 1999;96(21):11782–11786.

[62] Chiodo, F, Marradi, M, Calvo, J, Yuste, E, Penadés, S. Glycosystems in nanotechnology: gold glyconanoparticles as carrier for anti-HIV prodrugs. Beilstein Journal of Organic Chemistry. 2014;10(1):1339–1346.

[63] Sato, Y, Yoshioka, K, Murakami, T, Yoshimoto, S, Niwa, O. Design of biomolecular interface for detecting carbohydrate and lectin weak interactions. Langmuir. 2012;28 (3):1846–1851.

[64] Pandey, B, Bhattarai, JK, Pornsuriyasak, P, Fujikawa, K, Catania, R, Demchenko, AV, Stine, KJ. Square-wave voltammetry assays for glycoproteins on nanoporous gold. Journal of Electroanalytical Chemistry. 2014;717-718:47–60.

[65] Wang, Y, El-Boubbou, K, Kouyoumdjian, H, Sun, B, Huang, X, Zeng, X. Lipoic acid glycoconjugates, a new class of agents for controlling nonspecific adsorption of blood serum at biointerfaces for biosensor and biomedical applications. Langmuir. 2009;26(6):4119–4125.

[66] Kanda, V, Kitov, P, Bundle, DR, McDermott, MT. Surface plasmon resonance imaging measurements of the inhibition of Shiga-like toxin by synthetic multivalent inhibitors. Analytical Chemistry. 2005;77(23):7497–7504.

[67] Jung, C, Dannenberger, O, Xu, Y, Buck, M, Grunze, M. Self-assembled monolayers from organosulfur compounds: a comparison between sulfides, disulfides, and thiols. Langmuir. 1998;14(5):1103–1107.

[68] Svedhem, S, Öhberg, L, Borrelli, S, Valiokas, R, Andersson, M, Oscarson, S, Svensson, SC, et al. Synthesis and self-assembly of globotriose derivatives: a model system for studies of carbohydrate-protein interactions. Langmuir. 2002;18(7):2848–2858.

[69] Jain, PK, Huang, X, El-Sayed, IH, El-Sayed, MA. Noble metals on the nanoscale: optical and photothermal properties and some applications in imaging, sensing, biology, and medicine. Accounts of Chemical Research. 2008;41(12):1578–1586.

[70] Sharma, A, Bhattarai, JK, Alla, AJ, Demchenko, AV, Stine, KJ. Electrochemical annealing of nanoporous gold by application of cyclic potential sweeps. Nanotechnology. 2015;26 (8):085602.

[71] Schwartzberg, AM, Zhang, JZ. Novel optical properties and emerging applications of metal nanostructures†. The Journal of Physical Chemistry C. 2008;112(28):10323–10337.

[72] Rojo, J, Diaz, V, de la Fuente, JM, Segura, I, Barrientos, AG, Riese, HH, Bernad, A, et al. Gold glyconanoparticles as new tools in antiadhesive therapy. ChemBiochem. 2004;5 (3):291–297.

[73] Sunasee, R, Adokoh, CK, Darkwa, J, Narain, R. Therapeutic potential of carbohydrate-based polymeric and nanoparticle systems. Expert Opinion on Drug Delivery. 2014;11 (6):867–884.

[74] Bhattarai, JK, Sharma, A, Fujikawa, K, Demchenko, AV, Stine, KJ. Electrochemical synthesis of nanostructured gold film for the study of carbohydrate–lectin interactions using localized surface plasmon resonance spectroscopy. Carbohydrate Research. 2015;405: 55–65.

[75] Pandey, B, Tan, YH, Parameswar, AR, Pornsuriyasak, P, Demchenko, AV, Stine, KJ. Electrochemical characterization of globotriose-containing self-assembled monolayers on nanoporous gold and their binding of soybean agglutinin. Carbohydrate Research 2013;373:9–17.

[76] Alla, AJ, d' Andrea, FB, Bhattarai, JK, Cooper, JA, Tan, YH, Demchenko, AV, Stine, KJ. Selective capture of glycoproteins using lectin-modified nanoporous gold monolith. Journal of Chromatography A. 2015;1423:19–30.

[77] Ganesh, NV, Fujikawa, K, Tan, YH, Nigudkar, SS, Stine, KJ, Demchenko, AV. Surface-tethered iterative carbohydrate synthesis: a spacer study. The Journal of Organic Chemistry. 2013;78(14):6849–6857.

[78] Pornsuriyasak, P, Ranade, SC, Li, A, Parlato, MC, Sims, CR, Shulga, OV, Stine, KJ, et al. STICS: surface-tethered iterative carbohydrate synthesis. Chemical Communications (Cambridge, England). 2009(14):1834–1836.

[79] Dietrich, PM, Horlacher, T, Girard-Lauriault, PL, Gross, T, Lippitz, A, Min, H, Wirth, T, et al. Adlayers of dimannoside thiols on gold: surface chemical analysis. Langmuir. 2011;27(8):4808–4815.

[80] Belman, N, Jin, K, Golan, Y, Israelachvili, JN, Pesika, NS. Origin of the contact angle hysteresis of water on chemisorbed and physisorbed self-assembled monolayers. Langmuir. 2012;28(41):14609–14617.

[81] Hederos, M, Konradsson, P, Liedberg, B. Synthesis and self-assembly of galactose-terminated alkanethiols and their ability to resist proteins. Langmuir. 2005;21(7):2971–2980.

[82] Fyrner, T, Ederth, T, Aili, D, Liedberg, B, Konradsson, P. Synthesis of oligo(lactose)-based thiols and their self-assembly onto gold surfaces. Colloids and Surfaces B: Biointerfaces. 2013;105:187–193.

[83] Tromas, C, Eaton, P, Mimault, J, Rojo, J, Penadés, S. Structural characterization of self-assembled monolayers of neoglycoconjugates using atomic force microscopy. Langmuir. 2005;21(14):6142–6144.

[84] Yu, J-J, Nolting, B, Tan, YH, L, X, Grvay-Hague, J, Lu, G-y. Polyvalent interactions of HIV-gp120 protein and nanostructures of carbohydrate ligands. NanoBiotechnology. 2005;1 (2):201–210.

[85] Cheng, F, Shang, J, Ratner, DM. A versatile method for functionalizing surfaces with bioactive glycans. Bioconjugate Chemistry. 2011;22(1):50–57.

[86] Dhayal, M, Ratner, DM. XPS and SPR analysis of glycoarray surface density. Langmuir. 2009;25(4):2181–2187.

[87] Choi, I, Kim, YK, Min, DH, Lee, S, Yeo, WS. On-demand electrochemical activation of the click reaction on self-assembled monolayers on gold presenting masked acetylene groups. Journal of American Chemical Society 2011;133(42):16718–16721.

[88] Mukherjee, MD, Solanki, PR, Sumana, G, Manaka, T, Iwamoto, M, Malhotra, BD. Thiol modified chitosan self-assembled monolayer platform for nucleic acid biosensor. Applied Biochemistry and Biotechnology. 2014;174(3):1201–1213.

[89] Homola, J. Present and future of surface plasmon resonance biosensors. Analytical and Bioanalytical Chemistry. 2003;377(3):528–539.

[90] Schlick, KH, Cloninger, MJ. Inhibition binding studies of glycodendrimer-lectin interactions using surface plasmon resonance. Tetrahedron. 2010;66(29):5305–5310.

[91] Bhattarai, JK. Electrochemical Synthesis of Nanostructured Noble Metal Films for Biosensing. University of Missouri-St. Louis; 2014.

[92] Ringe, E, McMahon, JM, Sohn, K, Cobley, C, Xia, Y, Huang, J, Schatz, GC, et al. Unraveling the effects of size, composition, and substrate on the localized surface plasmon resonance frequencies of gold and silver nanocubes: a systematic single-particle approach. The Journal of Physical Chemistry C. 2010;114(29):12511–12516.

[93] Willets, KA, Van Duyne, RP. Localized surface plasmon resonance spectroscopy and sensing. Annual Review of Physical Chemistry 2007;58:267–297.

[94] Han, Y, Lupitskyy, R, Chou, T-M, Stafford, CM, Du, H, Sukhishvili, S. Effect of oxidation on surface-enhanced Raman scattering activity of silver nanoparticles: a quantitative correlation. Analytical Chemistry. 2011;83(15):5873–5880.

[95] Bellapadrona, G, Tesler, AB, Grunstein, D, Hossain, LH, Kikkeri, R, Seeberger, PH, Vaskevich, A, et al. Optimization of localized surface plasmon resonance transducers for studying carbohydrate-protein interactions. Analytical Chemistry. 2012;84(1):232–240.

[96] Morokoshi, S, Ohhori, K, Mizukami, K, Kitano, H. Sensing capabilities of colloidal gold modified with a self-assembled monolayer of a glucose-carrying polymer chain on a glass substrate. Langmuir. 2004;20(20):8897–8902.

[97] Kitano, H, Takahashi, Y, Mizukami, K, Matsuura, K. Kinetic study on the binding of lectin to mannose residues in a polymer brush. Colloids and Surfaces B: Biointerfaces. 2009;70(1):91–97.

[98] Mizukami, K, Takakura, H, Matsunaga, T, Kitano, H. Binding of Ricinus communis agglutinin to a galactose-carrying polymer brush on a colloidal gold monolayer. Colloids and Surfaces B: Biointerfaces. 2008;66(1):110–118.

[99] Kitano, H, Nakada, H, Mizukami, K. Interaction of wheat germ agglutinin with an N-acetylglucosamine-carrying telomer brush accumulated on a colloidal gold monolayer. Colloids and Surfaces B: Biointerfaces. 2008;61(1):17–24.

[100] Li, Y, Ma, B, Fan, Y, Kong, X, Li, J. Electrochemical and Raman studies of the biointeraction between Escherichia coli and mannose in polydiacetylene derivative supported on the self-assembled monolayers of octadecanethiol on a gold electrode. Analytical Chemistry. 2002;74(24):6349–6354.

[101] Lisdat, F, Schäfer, D. The use of electrochemical impedance spectroscopy for biosensing. Analytical and Bioanalytical Chemistry. 2008;391(5):1555–1567.

[102] Chang, B-Y, Park, S-M. Electrochemical impedance spectroscopy. Annual Review of Analytical Chemistry. 2010;3:207–229.

[103] Guo, X, Kulkarni, A, Doepke, A, Halsall, HB, Iyer, S, Heineman, WR. Carbohydrate-based label-free detection of Escherichia coli ORN 178 using electrochemical impedance spectroscopy. Analytical Chemistry. 2012;84(1):241–246.

[104] Shen, Z, Huang, M, Xiao, C, Zhang, Y, Zeng, X, Wang, PG. Nonlabeled quartz crystal microbalance biosensor for bacterial detection using carbohydrate and lectin recognitions. Analytical Chemistry. 2007;79(6):2312–2319.

[105] Mannelli, I, Minunni, M, Tombelli, S, Mascini, M. Quartz crystal microbalance (QCM) affinity biosensor for genetically modified organisms (GMOs) detection. Biosensors and Bioelectronics. 2003;18(2):129–140.

[106] Sharon, N, Lis, H. Lectins as cell recognition molecules. Science. 1989;246(4927):227–234.

[107] Lis, H, Sharon, N. Lectins: carbohydrate-specific proteins that mediate cellular recognition. Chemical Reviews. 1998;98(2):637–674.

[108] Naismith, JH, Field, RA. Structural basis of trimannoside recognition by concanavalin A. Journal of Biological Chemistry. 1996;271(2):972–976.

[109] Ambrosi, M, Cameron, NR, Davis, BG, Stolnik, S. Investigation of the interaction between peanut agglutinin and synthetic glycopolymeric multivalent ligands. Organic & Biomolecular Chemistry. 2005;3(8):1476–1480.

[110] Dam, TK, Brewer, CF. Thermodynamic studies of lectin-carbohydrate interactions by isothermal titration calorimetry. Chemical Reviews. 2002;102(2):387–430.

[111] Lin, C-C, Yeh, Y-C, Yang, C-Y, Chen, G-F, Chen, Y-C, Wu, Y-C, Chen, C-C. Quantitative analysis of multivalent interactions of carbohydrate-encapsulated gold nanoparticles with concanavalin A. Chemical Communications. 2003(23):2920–2921.

[112] Halkes, KM, Carvalho de Souza, A, Maljaars, CEP, Gerwig, GJ, Kamerling, JP. A facile method for the preparation of gold glyconanoparticles from free oligosaccharides and their applicability in carbohydrate-protein interaction studies. European Journal of Organic Chemistry. 2005;2005(17):3650–3659.

[113] Loaiza, OA, Lamas-Ardisana, PJ, Jubete, E, Ochoteco, E, Loinaz, I, Cabanero, G, Garcia, I, et al. Nanostructured disposable impedimetric sensors as tools for specific biomolecular interactions: sensitive recognition of concanavalin A. Analytical Chemistry. 2011;83 (8):2987–2995.

[114] Kaper, JB, Nataro, JP, Mobley, HL. Pathogenic *Escherichia coli*. Nature Reviews Microbiology. 2004;2(2):123–140.

[115] Marrs, CF, Zhang, L, Foxman, B. Escherichia coli mediated urinary tract infections: are there distinct uropathogenic E. coli (UPEC) pathotypes? FEMS Microbiology Letters. 2005;252(2):183–190.

[116] Martinez, JJ, Mulvey, MA, Schilling, JD, Pinkner, JS, Hultgren, SJ. Type 1 pilus-mediated bacterial invasion of bladder epithelial cells. The EMBO Journal. 2000;19(12):2803–2812.

[117] Liang, MN, Smith, SP, Metallo, SJ, Choi, IS, Prentiss, M, Whitesides, GM. Measuring the forces involved in polyvalent adhesion of uropathogenic Escherichia coli to mannose-presenting surfaces. Proceedings of the National Academy of Sciences of the United States of America. 2000;97(24):13092–13096.

[118] Lin, C-C, Yeh, Y-C, Yang, C-Y, Chen, C-L, Chen, G-F, Chen, C-C, Wu, Y-C. Selective binding of mannose-encapsulated gold nanoparticles to type 1 pili in Escherichia coli. Journal of the American Chemical Society. 2002;124(14):3508–3509.

Chapter 5

Melatonin: A Silent Regulator of the Glucose Homeostasis

Cristina Manuela Drăgoi, Andreea Letiția Arsene,
Cristina Elena Dinu-Pîrvu, Ion Bogdan Dumitrescu,
Daniela Elena Popa,
George T.A. Burcea-Dragomiroiu,
Denisa Ioana Udeanu, Olivia Carmen Timnea,
Bruno Ștefan Velescu and Alina Crenguța Nicolae

Additional information is available at the end of the chapter

http://dx.doi.org/10.5772/66625

Abstract

In the human organism, the circadian regulation of carbohydrates metabolism is essential for the glucose homeostasis and energy balance. Unbalances in glucose and insulin tissue and blood levels have been linked to a variety of metabolic disorders such as obesity, metabolic syndrome, cardiovascular diseases and type 2 diabetes. Melatonin, the pineal hormone, is the key mediator molecule for the integration between the cyclic environment and the circadian distribution of physiological and behavioral processes and for the optimization of energy balance and body weight regulation, events that are crucial for a healthy organism. This chapter reviews the interplay between melatonin modulatory physiological effects, glucose homeostasis and metabolic balance, from the endocrinology perspective. The tremendous effect of melatonin in the regulation of metabolic processes is observed from the chronobiology perspective, considering melatonin as a major synchronizer of the circadian internal order of the physiological processes involved in energy metabolism.

Keywords: melatonin, pineal gland, chronobiology, glucose homeostasis, metabolic syndrome

1. Introduction

Chronobiology depicts the temporal structure of biology. It integrates the rhythmic development and existence of the utmost majority of cells, comprising also cellular functions, as a fundamental property of the living matter, detectable at all levels of its organization, from the molecular level to rhythms in the integrity of the complex organisms. It regulates the species' biological clock, the circadian and seasonal biorhythms, synchronizing the existence of a cell within an organ, the functioning rationale of an organ within a system and the epigenetic temporal regulation of the whole living entity.

Synchronicity between external and internal circadian rhythms and harmony among molecular fluctuations within cells are essential for normal organ biology. Circadian clocks exist within multiple components of the metabolic, cardiovascular and immune systems, having the potential of affecting multiple cellular processes and, therefore, holding the promise of modulating various physiopathological aspects over the course of the 24 h cycle.

2. The pineal gland, a neuroendocrine transducer

Metabolic physiology undergoes diurnal variations, and serious pathologic events appear to be conditioned by the time of day. The suprachiasmatic nucleus imprints the control of circadian rhythms in peripheral tissues, by different neural and humoral signals, such as melatonin.

The molecular clock mechanism in mammals is currently understood as a transcriptional feedback loop involving several genes. The genes Clock and Bmal1 encode bHLH-PAS (basic helix-loop-helix) proteins that form the positive limb of the feedback circuit. The CLOCK:BMAL1 heterodimer initiates the transcription by binding to specific DNA elements, E-boxes (5′-CACGTG-3′) and E′-boxes (5′-CACGTT-3′) in the promoters of target genes. This set of activated genes includes members of the negative limb of the feedback loop including the PER (PER1 and PER2) and CRY (CRY1 and CRY2) genes. The resulting PER and CRY proteins dimerize and inhibit further CLOCK:BMAL1 transcriptional activity allowing the cycle to repeat from a level of low transcriptional activity. Thus, cellular metabolism may prove to play an important role in regulating the transcriptional state and therefore the phase of the clock. Degradation of the negative limb proteins PER and CRY is required to terminate the repression phase and restart a new cycle of transcription. The transcriptional feedback loop described above can be observed not only in the SCN, but also in nearly every mammalian cell. If viewed at the single-cell level, the molecular clockwork of transcription and translation can be observed as autonomous single-cell oscillators [1].

Melatonin (*N*-acetyl-5-methoxytryptamine) is synthesized by multiple tissues in the body, but the pineal gland is the major contributor to circulating melatonin concentration, as pinealectomy abolishes detectable melatonin in the blood.

In young- and middle-aged people, pineal melatonin is secreted based on a circadian pattern, with a high rhythm amplitude and a considerable nocturnal maximum.

Melatonin from extrapineal sites often oscillates with considerably lower amplitudes. According to current knowledge, some of the extrapineal sources are of particular importance, either in quantitative terms, such as the gastrointestinal tract or, with regard to functional aspects, some areas of the central nervous system (CNS) and several leukocytes. The physiological significance of other sites of melatonin biosynthesis is, at the moment, uncertain. Melatonin is secreted in small amounts from most of the extrapineal sites or only under specific conditions, for example, the postprandial release from the gastrointestinal tract, during which relatively high quantities can enter the circulation, being chronobiologically rather irrelevant. Thus, melatonin is not only a pineal hormone but also has additional functions as a local tissue factor and leukocyte-derived cell hormone with paracrine and autocrine actions [2].

The pineal gland or the epiphysis weighs about 150 mg, and it is located in the posterior part of the third cerebral ventricle. The pineal gland of mammals is a homogeneous tissue containing pinealocytes, glial cells, phagocytic cells and neurons. The pineal gland is innervated by nervous fibers of different origins. The gland was considered a vestigial organ until 1950s. Its position in the center of the brain and its presence in other species of vertebrates indicate its evolution in the evolutionary cerebral system of the humans, based on its absolute necessity on the overall organism development and important particular functions.

In the current scientific frame, it is generally acknowledged that the pineal organ is a neuroendocrine transducer. The pinealocytes, the main secretory pineal parenchymal cells, are fotoneuroendocrine cells, phylogenetically derived from primary sensory cells, having neural embryological origin. They primary respond to nervous-photic modulated stimulation and secondary to hormones in target organs. The endocrine secretory function is directly dependent on sympathetic innervation, the pinealocytes translating the nervous information in endocrine information.

In terms of phylogenetic evolution, the pinealocyte functions as a fotoneuroendocrine neuron. On some species of amphibians, reptiles and birds, the epiphysis is also called "parietal eye" or third eye because it has the form of a rudimentary eye fitted with a lens and a retina. Therefore, it is considered to be a vestigial of a sensory organ, functional in primitive vertebrates. Being directly affected by the light absorbed by the eye, the pineal gland regulates the sleep-wake states, the menstruation and the reproduction, the hibernation and the seasonal migration and other "instinctive" behaviors.

The role of the pineal gland, as an integral component of the brain, evolved on the phylogenetic scale from photoreceptor organ (fish, amphibians, reptiles), to neuroendocrine modulator of brain functions, in mammals and humans, thus having a role in the adaptation of reproductive conditions environment in some mammals (particularly rodents) and to a major role in the modulation of brain excitability in relation to the external environmental conditions in humans.

Nowadays, melatonin, the major pineal hormone, is considered the master hormone, regulating all other hormones within the human organism.

The main regulatory pathway is a complex one called the retinoic-hypothalamic-pineal axis that ends with the sympathetic transmission of the pineal parenchymal. The pineal gland receives neuronal transmissions of central and parasympathetic origin. These pineal nervous endings contain a great variety of neurotransmitters. The rhythm of melatonin synthesis depends on three interrelated factors: the endogenous circadian oscillator, located in the suprachiasmatic nucleus (SCN), the light/dark and day/night cycles that synchronize the endogenous oscillator and the light that dramatically inhibits the synthesis of melatonin [3].

In conclusion, the pineal gland is a connection point between various neural transmissions, its activity being under the high-fidelity control of the hypothalamic clock, the temporal message being delivered to the pineal gland via a polysynaptic path. However, new neuroanatomic and immunocytochemical proofs changed the concept according to which the pineal gland is innervated only by the sympathetic nervous system, presently, being unanimously accepted to be the target of several neurotransmitters of different origins.

3. Melatonin, the universal synchronizer

Melatonin biosynthesis at the level of pinealocytes occurs and is initiated by the absorption of tryptophan from the blood. Increased daytime tryptophan concentration at the pineal level precedes increased serum free and total tryptophan, suggesting that the essential amino acid is captured by the pineal against a concentration gradient. Once arrived in the pinealocyte, the major part of tryptophan is used for the synthesis of indole derivatives and the rest for protein synthesis (**Figure 1**).

The transformation of tryptophan into serotonin occurs in two stages: first, the hydroxylation into 5-hydroxytryptophan under the enzymatic action of tryptophan hydroxylase, this being a limiting step of the synthesis. The enzyme activity requires the presence of oxygen, tetrahydrobiopterin, $NADPH^+$ and a metal, iron or copper. Second, the decarboxylation of 5-hydroxytryptophan into serotonin by the action of L-aromatic amino acid decarboxylase, in the presence of pyridoxal phosphate (PLP).

The transformation of serotonin into melatonin also includes two stages: the acetylation of $-NH_2$ group by the *N*-acetyltransferase (NAT) enzyme to form *N*-acetyl-serotonin, and thereafter, the methylation of –OH group in position 5 by hydroxyl-indole-*O*-methyltransferase (HIOMT), an enzyme that catalyzes the transfer of methyl group from *S*-adenosyl methionine. This final step results in acetyl-5-methoxytryptamine or melatonin synthesis. The two enzymes, NAT and HIOMT, specific for this synthesis pathway, have different profiles of activity. The NAT enzyme is a limiting enzyme of reaction: it is subject to many mechanisms of transcriptional and/or posttranscriptional regulation depending on the species, which allows it to be active only during darkness. The NAT activity is strongly regulated by circadian alternation light/dark, contrary to HIOMT enzyme showing a constitutive activity along the nictemeral cycle. For example, in rats and humans, a short exposure to bright light during the dark period causes an inhibition of the NAT activity for the next 15 min [3].

The enzymes responsible for the melatonin production can be modulated in a circadian manner, in order to imprint its nictemeral synthesis cycle.

Extrapineal melatonin synthesis in mammals has been reported in the retina, Harderian glands, gastrointestinal tract and pancreas. In humans, melatonin has been reported outside

http://dx.doi.org/10.5772/66625

the pineal gland in the follicles, the lining of the intestinal appendix, platelets and red blood cells. Melatonin is also produced by numerous nonendocrine cells, as is the case of the immune cells. In conclusion, while the pineal gland quantitatively accounts for most of the circulating melatonin, substantial local synthesis also occurs in retinal and peripheral tissues such as the gastrointestinal tract [4].

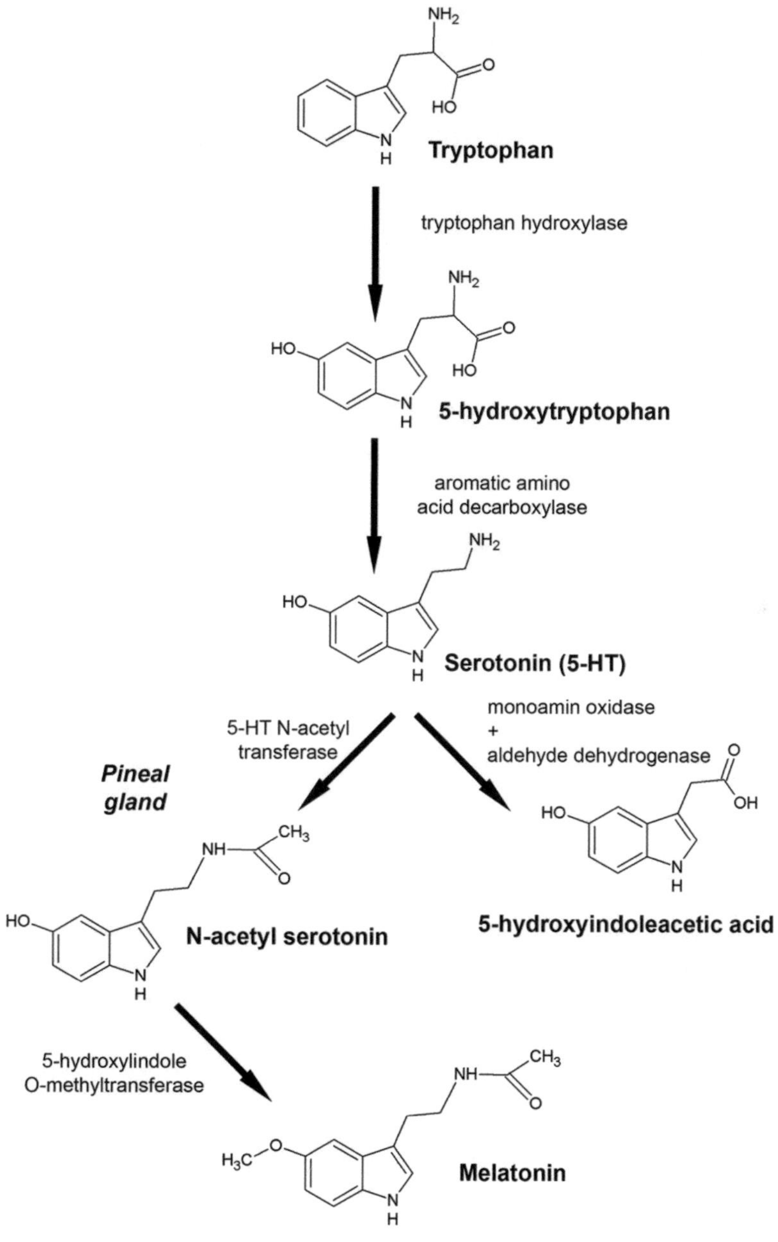

Figure 1. Melatonin biosynthesis pathway.

Melatonin appears to be secreted by the pineal gland in circulation by simple diffusion, because it is highly soluble in the cell membrane lipoproteins. Furthermore, melatonin may have effects on the pineal gland itself, as there are specific receptors for it, at this level. The cells of the suprachiasmatic nucleus also possess receptors for melatonin. In conclusion, melatonin has an inhibitory effect on the activity of the suprachiasmatic nucleus. So, melatonin is self-regulating its own synthesis. However, melatonin secreted into the bloodstream will send to all central and peripheral structures that possess receptors or melatonergic sites this information regarding the photoperiodicity, allowing the organism a physiological adaptation to alternations day/night or to the seasonal ones.

The circadian pacemaker within the suprachiasmatic nucleus triggers the pineal gland to produce high melatonin concentrations at night. There is a photic synchronized endogenous circadian biorhythm, allowing the maximum human melatonin production at night, between midnight and 3 a.m., and the serotonin during the day (**Figure 2a**). Initiating the melatonin synthesis in humans occurs between 9 and 11 p.m. and lasts for about 8–9 h in adults, these parameters being fairly constant from day to day. The daily and seasonal melatonin rhythms are involved in time of day and time of year signaling, and it is for this reason that they are considered to serve as a bioclock and biocalendar.

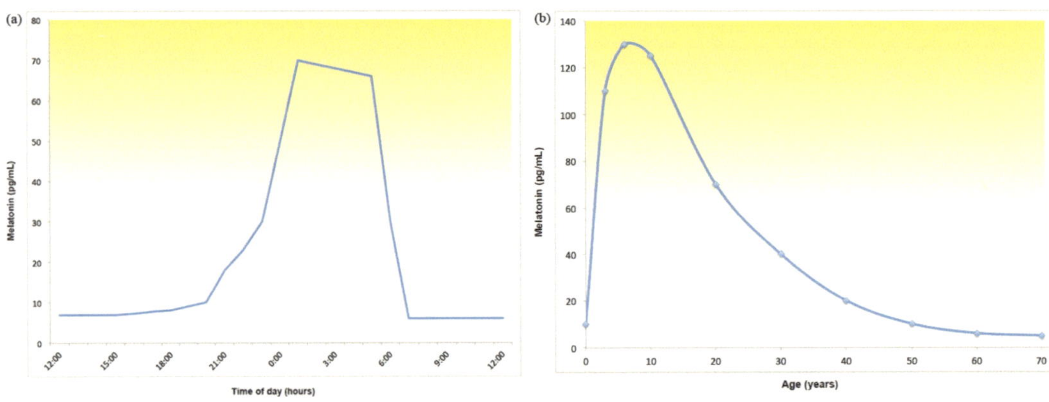

Figure 2. The daily (a) and lifetime (b) melatonin biosynthesis fluctuations.

In humans, it is required an intensity of light greater than in other mammals, over 1500 lux, for disrupting this synthesis biorhythm by external light. In all studied mammals, those with activity at day and at night, the nocturnal melatonin production is maximal and prevailing overnight.

In humans, blood levels of melatonin have a particular dynamics from birth (**Figure 2b**). The pineal synthesis begins in infants over 3 months old, to older age, and the time of puberty is very controversial whether it represents or not a particular step on the downward slope. It is of real interest the fact that melatonin synthesis decreases with the age of humans and has abnormal low levels in a series of age-related pathological disorders, as it is the case of cardiovascular and metabolic disease. Melatonin rhythmic profile has many implications in pathophysiological processes as inflammation, oxidative stress, hypertension and metabolic syndrome.

The half-life of blood melatonin is under 30 min, and the metabolic clearance is 630 mL/min in healthy men. Melatonin is a lipophilic substance metabolized in several compounds both in the liver (**Figure 3a**) and in the central nervous system (**Figure 3b**) [3, 5].

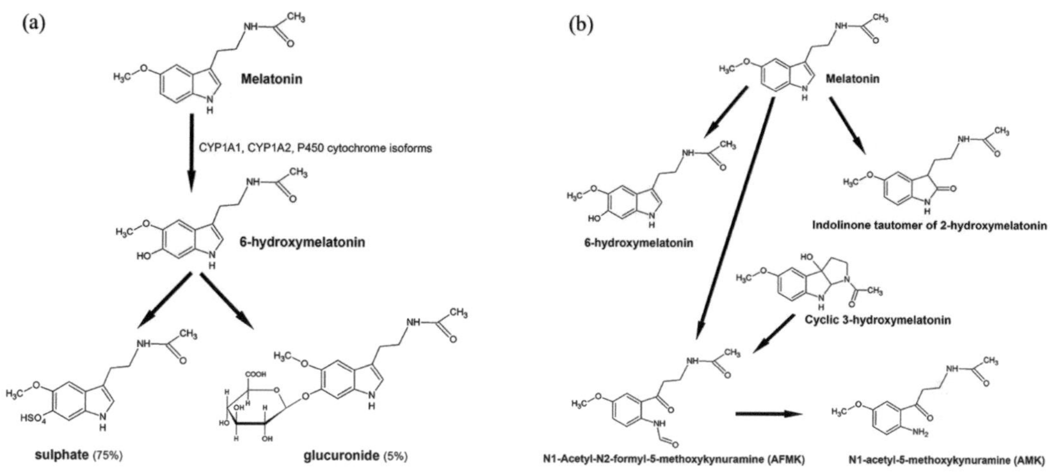

Figure 3. Melatonin hepatic (a) and cerebral (b) metabolism.

Circulating melatonin is catabolized in humans, under the action of hepatic microsomal hydroxylases in *N*-acetylserotonin, which is biologically inactive, and 6-hydroxymelatonin, which is urinary eliminated, as a sulfate (75%) or glucuronide (5%) conjugate. The rest of melatonin is eliminated under three forms: the native form (below 1%), as 5-metoxy indol-acetic acid (0.5%) or catabolized as *N*-acetyl-5-metoxy kynurenamine via *N*-formyl-5-metoxy kynurenamine in the central nervous system (15%).

In the brain, melatonin is metabolized in several compounds of which 5-metoxytriptamine is involved in the dreaming process and it will subsequently metabolize N, *N*-dimethyl-5-methoxytryptamine and other tryptamines.

Approximately 60–70% of the circulating melatonin in the bloodstream is bound to albumin, and about 30% is found in free state, this fraction being the one, that crosses the blood brain barrier [6, 7].

4. Melatonin's pleiotropic effects and the metabolic epigenetic regulation

Considered as nature's most versatile biological signaling and multitasking molecule, melatonin is a highly conserved molecule found in almost all types of organisms. Melatonin has several functions ranging from coordination of circadian activity, which is generally considered as a sleep-promoting effect, and melatonin administration induces hypothermic effects and heat loss via the distal skin regions, and stabilizes sleep-wake cycles [8].

Melatonin demonstrates properties of a powerful antioxidant, at sufficiently high concentrations as a direct radical scavenger, but, at lower, near-physiological levels, as a regulator of redox-relevant enzymes, suppressor of prooxidant excitatory and inflammatory processes and as a mitochondrial modulator [2]. Melatonin also acts on bone metabolism, activating its MT1 and MT2 specific receptors in an autocrine or paracrine fashion, near the target tissues. Melatonin may exert vasodilatatory (MT2) or vasoconstriction (MT1) effects, depending on the receptor type or the target cell, and it can also down-regulate the cortisol secretion [9].

Melatonin has a great influence on diabetes and associated metabolic unbalances by regulating insulin secretion, and also by scavenging reactive oxygen species, the pancreatic β-cells being highly susceptible to oxidative stress, and possessing only low-antioxidative potential. On the other hand, in several genetic studies, human MT2 receptor polymorphisms have been described as being causally linked to an elevated risk of developing type 2 diabetes [4].

5. Chronobiology, metabolic control and disease modulation by melatonin

It is universally accepted that social and industrial pressures, such as shift work, which opposes the physiological temporal circadian order, may be factors determining the occurrence or the development of chronic illnesses, such as metabolic disorders. In many disease states as diabetes mellitus and hypertension, neurohumoral circadian rhythms are chronically impaired and result in dyssynchrony of cellular order in different tissues and between the organism and the environment.

Diabetes mellitus is associated with a phase shift in the cardiac circadian clock. Shift workers have an increased incidence of cardiovascular disease, which might be related to alterations in cardiovascular and metabolic intracellular circadian clock function [5, 10].

The cardiovascular system actually exhibits significant daily variation regarding physiological, pathophysiological and molecular processes. Diurnal variations also affect gene and protein expression. Circadian clocks exist in cardiomyocytes, vascular smooth muscle cells and endothelial cells.

At molecular level, a complex interplay occurs between environmental influences and intrinsic mechanisms, which contributes to change in metabolic functions over the 24 h period.

Loss of synchronization occurs when there are changes in feeding or sleep patterns and during exposure to light at abnormal times, at night, considering this phenomenon as "light-at-night pollution." Such dyssynchronization is seen in patients with hypertension, diabetes mellitus, obesity and shift workers, in whom there is an elevated risk of cardiovascular disease [5, 11].

Melatonin is the key mediator molecule in the in vivo scenario, mediating the integration between the cyclic environment and the circadian distribution of physiological and behavioral processes.

The relation between pineal gland, melatonin and energy metabolism was initially studied in both humans and rodents. Over 70 years ago, one of the first references to the functional connection between the pineal gland and carbohydrates metabolism was made by the Romanian researcher Constantin I. Parhon, followed by his coworkers, endocrinologists Milcu and Nanu. They conducted animal experiments on "pinealin," a pineal peptide, described as having metabolic effects similar to insulin, displaying hypoglycemic, anabolic, anticholesterolemic and glomerulotrophic characteristics. In the following years, a controversial discussion was carried out in many publications on the importance of pineal extracts on glucose metabolism. Even after the isolation and identification of the molecular structure of melatonin, by Aron Lerner and colleagues, this discussion continued [12].

6. The functional synergism between melatonin and insulin

The scientific literature states that first experimental injections with pineal extracts led to hypoglycemia, increased glucose tolerance, and hepatic and muscular glycogenesis. The metabolic disruption caused by the absence of melatonin in the pinealectomized animals was characterized as a diabetogenic syndrome depicted by glucose intolerance and insulin resistance, both expressed peripherally: hepatic, adipose, and skeletal muscle, and centrally, at the level of the hypothalamus. This pathological picture can be reverted by melatonin replacement therapy or restricted feeding.

In addition to this dramatic finding, insulin resistance, glucose intolerance, and several other metabolic disorders can be seen in some physiological or pathological states associated with reductions in blood melatonin levels, as aging, diabetes, shift work, and environmental illumination during the night and the so-called phenomenon of light pollution. An adequate melatonin replacement therapy alleviates most of these alterations.

Emphasizing the functional synergism between melatonin and insulin, it is considered that the pinealectomy-induced insulin resistance and glucose intolerance are related to the mechanistic consequences of the depletion of melatonin, perceived at the molecular level as a deficiency in the insulin-signaling pathway and reduction in GLUT4 gene expression and protein content. It was shown that melatonin, acting through MT1 membrane receptors, induces rapid tyrosine phosphorylation and activation of the tyrosine kinase beta-subunit of the insulin receptor, succeeding to overcome several intracellular transduction steps of the insulin-signaling pathway [6, 7].

7. Melatonin effects on adipocytes

The melatonin-insulin synergism was described in an in vitro experiment which supposed the incubation of isolated visceral white adipocytes with melatonin, the peripheral function of insulin being potentiated by the action of melatonin, and, in addition, this was the first evidence of a direct action of melatonin on adipocytes [6].

All in all, this was a proof that the adipose tissue is a peripheral target of melatonin for the regulation of the overall metabolism. Similarly, it was demonstrated that melatonin activation of MT2 receptors in human adipocytes modulates glucose uptake by these cells. Considering the adipose tissue physiology, it was possible to document synergistic effect of melatonin on several other insulin actions in addition to glucose uptake: insulin-induced leptin synthesis and release in isolated adipocytes is potentiated by the MT1-mediated melatonin action, and melatonin regulates other aspects of adipocyte biology that influence energy metabolism, lipidemia and body weight, as lipolysis, lipogenesis, adipocyte differentiation and fatty acids uptake [6].

Melatonin also exerts different effects on the carbohydrates metabolism, considering various targets: it stimulates glucose uptake in muscle cells by phosphorylation of insulin receptor substrate-1 through MT2 signaling, MT2 receptors are expressed in hepatocytes, and melatonin therapy elevates glucose release from the liver [9].

Another major site of melatonin's action in reference to the regulation of energy metabolism is the pancreatic islets where it influences insulin and glucagon synthesis and release.

Molecular and immunocytochemical studies confirmed the presence of melatonin receptors MT1 and MT2 in the islets of Langerhans and also in human pancreatic tissue [13]. MT1 and/or MT2-mediated melatonin action decreases glucose-stimulated insulin secretion in isolated rat pancreatic islets and rat insulinoma beta-cells.

Melatonin influences exocytosis of insulin by β-cells as concluded from experiments via nonhydrolyzable guanosine-5′-trisphosphate (GTP) analog and luzindole, a melatonin antagonist, both of which inhibit the melatonin action on secretion of insulin from neonatal rat islets.

The intracellular signal transduction pathways of the pancreatic β-cell influenced by melatonin via MT1 and MT2 membrane receptors include cAMP, cGMP and IP3 signaling pathways. The activation of these receptors inhibits glucose- and forskolin-induced insulin secretion, showing that melatonin acts by inhibiting the adenylate cyclase/cAMP system and reducing the content of PKA.

The pineal indolamine induces IGF-1 receptor phosphorylation, which participates in the integrity of islet cells. Moreover, it has been demonstrated, as well, that melatonin stimulated glucagon synthesis and secretion.

Above all considerations, these actions of melatonin are required to build the circadian profile of insulin secretion, synchronizing the pancreatic metabolic rhythms with the circadian rhythm of the activity/rest profile. And it should be also noted that insulin is able to regulate pineal melatonin synthesis by potentiating norepinephrine-stimulated melatonin production.

Interestingly, the association between melatonin and type 2 diabetes could be based on the observation that insulin secretion is inversely proportional to plasma melatonin concentration. These two hormones, melatonin and insulin, exhibit a circadian rhythm, but there is negative correlation between their synthesis dynamics.

Decreased abnormally regulated melatonin levels have been related to diabetes, which suggests that the melatonin signal is critical for glucose homeostasis. In patients with type 2 diabetes, gluconeogenesis and endogenous glucose production exhibit circadian rhythms that impose fasting high blood glucose and do not exist in healthy humans.

Melatonin inhibits glucose-mediated release of insulin from pancreatic cells emphasizing its activity in the function of insulin. Suppression of melatonin secretion by nocturnal light exposure could be a trigger for type 2 diabetes development (**Figure 4**) [9].

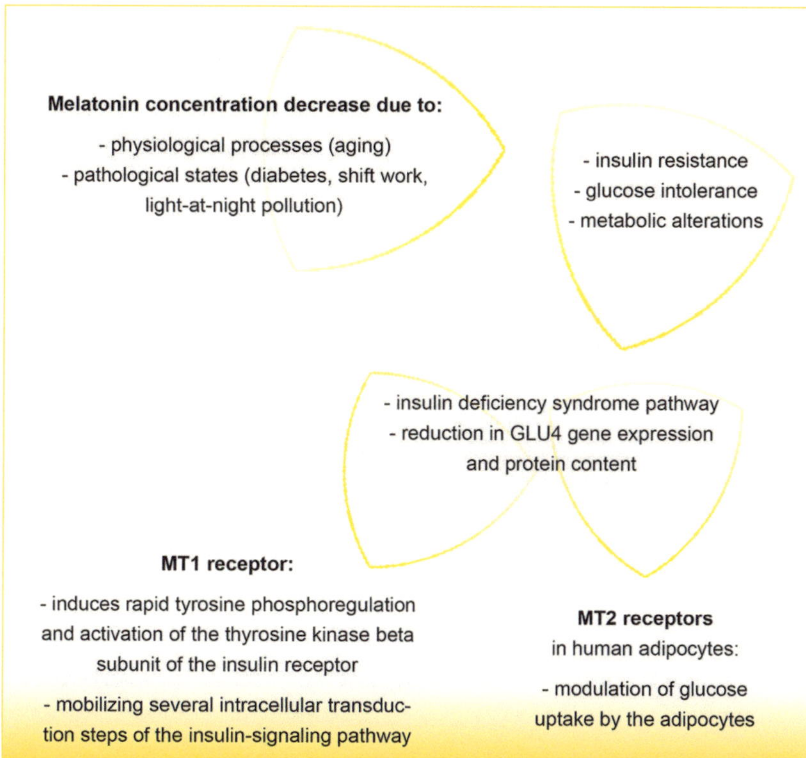

Figure 4. Melatonin depletion induced pathological consequences.

As an addition to the importance of melatonin on the regulatory processes in energy metabolism, it was recently demonstrated that the intrauterine metabolic programming is completely disturbed by the deficiency of melatonin in the pregnant mother. The adult child of a melatonin unpaired mother presents glucose intolerance, insulin resistance and a serious delay in the glucose-induced insulin secretion by isolated pancreatic islets [6, 13, 14].

This in once more a clear evidence that melatonin has a crucial role in the metabolic epigenetic regulation of a healthy organism that undergoes vicious trials and environmental demands that are, to a certain extent, meant to test the physiological robustness.

Robustness is one of the fundamental organizational principles of biological systems, this being the major characteristic involved in their adaptation, survival and reproduction. Metabolic diseases are considered a breakdown in the robustness in biological systems, the continuous maintenance of physiological functions being of extreme importance, despite external and internal disturbances [4].

Melatonin is a powerful chronobiotic, regulating the daily metabolic processes so that the activity/feeding phase of the day is assimilated with high insulin sensitivity, and the rest/fasting is synchronized to the insulin-resistant metabolic phase of the day. Melatonin is the key mediator molecule for the nictemeral integration of physiological and behavioral processes and for the modulation of energy balance and body weight regulation, all crucial for a healthy life [6].

The hypothalamus controls a great variety of physiological processes, including sleep/wake cycles, sexual behavior and reproduction, and metabolic control such as thermoregulation, glucose metabolism, lipid metabolism, energy intake/expenditure, and food and water intake, all these functions following circadian rhythms.

The hypothalamus identifies nutrients such as glucose and lipids, and via a specialized area of the blood brain barrier in the arcuate nucleus it also detects circulating metabolic hormones such as leptin, insulin, thyroid hormone, adiponectin and ghrelin. Researchers showed that SCN lesions abolished the daily rhythms in plasma concentrations of glucose and insulin and revealed the existence of a pronounced day/night difference in the response to 2-deoxy glucose, a glucose-utilization inhibitor, proving that the SCN is involved in the daily rhythm of glucose metabolism. SCN-lesioned rats do not have a rhythm in food intake either, so it should not be excluded an indirect effect of the lack of a feeding rhythm on glucose metabolism [11].

The chronobiotic nature of energy balance and energy metabolism is depicted by the two separated phases that exist during a 24 h period. The first one is characterized by energy harvesting and eating that results in energy intake, utilization, and storage, a period associated with high central and peripheral sensitivity to insulin and high glucose tolerance, elevated insulin secretion, high glucose uptake by the insulin-sensitive tissues, glycogen synthesis and hepatic and muscular glycolysis, blockade of hepatic gluconeogenesis, and increased adipose tissue lipogenesis and adiponectin production.

In opposition, the second one, the rest/sleep phase of the day, is characterized by fasting periods that require the use of stored energy for maintaining the cellular homeostasis, exhibiting insulin resistance, accentuated hepatic gluconeogenesis and glycogenolysis, adipose tissue lipolysis, and leptin secretion [6].

Other hormones exerting modulatory effects on cellular metabolism, as glucocorticoids, growth hormone and catecholamines, present circadian rhythmic fluctuations in their secretion and activity. Melatonin, as an orchestrating factor in the circadian organization of the metabolic processes, prepares and modifies the central and peripheral metabolic tissues in order to respond to these hormones.

The antiobesogenic effect of melatonin is, in part, a result of its regulatory role on the balance of energy, acting mainly on the regulation of the energy mobilization from the stores and in energy expenditure. It was demonstrated that in healthy young animals, melatonin supplementation therapy reduces long-term body weight gain and the size of the visceral fat deposits, effects independent on the reduction in food intake. The same antiobesity protective effect of melatonin was seen in experiments of diet-induced obesity [6, 15, 16].

The adequate supplementation of melatonin lowers body weight and regulates body weight gain as well as the intra-abdominal visceral fat accumulation, as a result of the reestablishment of the circadian distribution of energy metabolism, the insulin signaling pathway reinstatement, the consequent disappearance of insulin resistance and glucose intolerance and, most importantly, the accentuation of the energy expenditure over the energy intake.

8. Conclusions

Melatonin is the key modulatory molecule responsible for the integration between the cyclic environment and the circadian distribution of physiological and behavioral processes, assuring a healthy metabolism and the optimization of energy balance and body weight regulation.

The circadian system may be a tractable target for decreasing the prevalence of metabolic disturbances. Melatonin acts by potentiating central and peripheral insulin action either due to regulation of GLUT4 expression or triggering the insulin-signaling pathway. Melatonin is responsible for maintaining an adequate energy balance mainly by regulating energy flow and the energy expenditure through the activation of brown adipose tissue. It also assures the metabolic processes respect the nictemeral physiology of the two major phases existent during 24 h, the activity/wakefulness/feeding state versus the rest/sleep/fasting phase.

The decline in melatonin synthesis, during physiological processes as aging, or pathology associated events as shift-work or illuminated environments during the night, induces insulin resistance, glucose intolerance, sleep disturbances and metabolic circadian disorganization, depicting a state of chronodisruption and metabolic imbalances, aggravating the general health state.

The present evidence that melatonin induces insulin secretion and can improve β-cell function certify melatonin supplementation as an accurate therapeutic approach for glucose homeostasis reestablishment. The available scientific proofs support the suggestion that melatonin replacement therapy, if adequately carried out, in terms of dose, formulation and time of the day of administration, might contribute to maintaining optimal blood glucose levels in diabetic patients and restore the chronobiotic order for achieving a more robust healthy state of the organism.

Acknowledgements

This study was partially funded by the University of Medicine and Pharmacy "Carol Davila," through the research project "Young Researchers" number 33886/11.11.2014.

Author details

Cristina Manuela Drăgoi[1], Andreea Letiţia Arsene[2], Cristina Elena Dinu-Pîrvu[3], Ion Bogdan Dumitrescu[4], Daniela Elena Popa[5], George T.A. Burcea-Dragomiroiu[5], Denisa Ioana Udeanu[6], Olivia Carmen Timnea[7], Bruno Ştefan Velescu[8] and Alina Crenguţa Nicolae[1*]

*Address all correspondence to: alinanicolae29@gmail.com

1 Department of Biochemistry, Faculty of Pharmacy, University of Medicine and Pharmacy "Carol Davila," Bucharest, Romania

2 Department of General and Pharmaceutical Microbiology, Faculty of Pharmacy, University of Medicine and Pharmacy "Carol Davila," Bucharest, Romania

3 Department of Physical and Colloidal Chemistry, Faculty of Pharmacy, University of Medicine and Pharmacy "Carol Davila," Bucharest, Romania

4 Department of Pharmaceutical Physics and Informatics, Faculty of Pharmacy, University of Medicine and Pharmacy "Carol Davila," Bucharest, Romania

5 Department of Drug Control, Faculty of Pharmacy, University of Medicine and Pharmacy "Carol Davila," Bucharest, Romania

6 Department of Clinical Laboratory and Food Safety, Faculty of Pharmacy, University of Medicine and Pharmacy "Carol Davila," Bucharest, Romania

7 Department of Medical Sciences, Faculty of Physical Education and Sports, Ecologic University of Bucharest, Romania

8 Department of Pharmacology and Clinical Pharmacy, Faculty of Pharmacy, University of Medicine and Pharmacy "Carol Davila", Bucharest, Romania

References

[1] Johnston J.D., Skene D.J. Regulation of mammalian neuroendocrine physiology and rhy thms by melatonin. Journal of Endocrinology. 2015;JOE-15-0119. doi:10.1530/JOE-15-0119

[2] Hardeland R. Neurobiology, pathophysiology, and treatment of melatonin deficiency and dysfunction. The Scientific World Journal. 2012;640389. doi:10.1100/2012/640389

[3] Drăgoi C.M. Tryptophan, serotonin, melatonin-the spectacular triad: physiological, pathological and therapeutic implications of some bio-compounds with indolic structure. LAP Lambert Academic Publishing, Germany; 2013.

[4] Espino J., Pariente A.J., Rodríguez A.B. Role of melatonin on diabetes-related metabolic disorders. World Journal of Diabetes. 2011 June 15;2(6):82–91. doi:10.4239/wjd.v2.i6.82

[5] Dominguez-Rodriguez A., Abreu-Gonzalez P., Sanchez-Sanchez J.J., Kaski J.C., Reiter R.J. Melatonin and circadian biology in human cardiovascular disease. Journal of Pineal Research. 2010;49:14–22. doi:10.1111/j.1600-079X.2010.00773.x

[6] Cipolla-Neto J., Amaral F.G., Afeche S.C., Tan D.X., Reiter R.J. Melatonin, energy metabolism, and obesity: a review. Journal of Pineal Research. 2014;56:371–381. doi:10.1111/jpi.12137

[7] Drăgoi C.M. Biochemical mechanisms of action of some bio-compounds with indolic structure [thesis]. Bucharest; 2012.

[8] Nduhirabandi F., du Toit E., Lochner A. Melatonin and the metabolic syndrome: a tool for effective therapy in obesity-associated abnormalities? Acta Physiologica. Forthcoming. doi:10.1111/j.1748-1716.2012.02410.x

[9] Cardinali D.P., Hardeland R. Inflammaging, metabolic syndrome and melatonin: a call for treatment studies. Neuroendocrinology. Forthcoming. doi:10.1159/000446543

[10] Kalsbeek A., la Fleur S., Fliers E. Circadian control of glucose metabolism. Molecular Metabolism. 2014;3(4):372–383.

[11] Peschke E., Bähr I., Mühlbauer E. Melatonin and pancreatic islets: interrelationships between melatonin, insulin and glucagon. International Journal of Molecular Sciences. 2013;14(4):6981–7015. doi:10.3390/ijms14046981

[12] Owino S., Contreras-Alcantara S., Baba K., Tosini G. Melatonin signaling controls the daily rhythm in blood glucose levels independent of peripheral clocks. PLos One. 2016;11(1). doi:10.1371/journal.pone.0148214

[13] Sharma S., Singh H., Ahmad N., Mishra P., Tiwari A. The role of melatonin in diabetes: therapeutic implications. Archives of Endocrinology and Metabolism. 2015;59(5). doi:10.1590/2359-3997000000098

[14] Nicolae A.C., Drăgoi C.M., Ceaușu I., Poalelungi C., Iliescu D., Arsene A.L. Clinical implications of the indolergic system and oxidative stress in physiological gestational homeostasis. Farmacia. 2015;63(1):46–51.

[15] Drăgoi C.M., Nicolae A.C., Grigore C., Dinu-Pîrvu C., Arsene A.L. Characteristics of glucose homeostasis and lipidic profile in a hamster metabolic syndrome model, after the co-administration of melatonin and irbesartan in a multiparticulate pharmaceutical formation. In: March; Bucharest. Book of abstracts: ASRMN; 2016. p. 25.

[16] Suzana E.V., Diana L.D., Adrian E.R., Vlad Z., Diana M.C., Andreea L.A., Cristina M.D., Alina C.N., Leon Z., Torsten S., Ana-Maria Z. Behavioral and molecular effects of prenatal continuous light exposure in the adult rat. Brain Research. 2016;1650:51–59.

Chapter 6

Glass Transition of Ultrathin Sugar Films Probed by X-Ray Reflectivity

Shigesaburo Ogawa and Isao Takahashi

Additional information is available at the end of the chapter

http://dx.doi.org/10.5772/66432

Abstract

Besides being the main types of carbohydrate in food, sugars are a representative protectant in biopharmaceutical formulations. To identify the protection mechanism, researchers have extensively investigated the bulk physicochemical properties of sugars. However, whereas the glass transition of sugar has been widely studied and debated, the physicochemical properties of sugar molecules in confined circumstances such as nanometer thick films remain largely unknown. In this chapter, we introduce an experimental procedure for analyzing the glass transition of sugars in ultrathin films. The analysis is based on X-ray reflectivity (XRR) analysis, which has been often applied in glass transition studies of polymer films, but never in sugar media.

Keywords: sugar, ultrathin film, thickness, X-ray reflectivity

1. Introduction

Amorphous sugars form a glassy matrix under far from equilibrium conditions (i.e., supercooling or supersaturation) [1–3]. For instance, sugar glass is obtained by cooling the sugar melt to temperatures far below the melting point (T_m) [1] or by greatly concentrating the sugar solution (i.e., aqueous sugar solution) by evaporation, or by freezing the solvent [2, 3]. The glassy sugar matrix restricts the mobility of incorporated biological molecules and the diffusion of active chemicals, effectively preventing the aggregation and gelation of the biological molecules. Understanding the features of such "sugar glasses," which are already greatly exploited in the food, biology, biochemistry, and pharmaceutical industries [4–10], is pivotal to further development of those diverse fields. Among the numerous sugar compounds, nonreduced oligosaccharides trehalose (Tre) and sucrose (Suc) are generally accepted as representative stabilizers of biopharmaceutical materials. The structures of these sugars are depicted in **Figure 1a** and **b**, respectively.

The specific heat and coefficient of expansion of a sugar melt abruptly change across the glass transition [1]. In the glassy state, both values are much closer to the crystalline values than the liquid-state values. The glass transitions of bulk sugar-water mixtures at various concentrations are commonly studied by differential scanning calorimetry (DSC), which detects the glass transition as a change in heat capacity. However, DSC thermograms often contain a problematic overshoot peak caused by a macroscopic dynamic process accompanying structural relaxation at the glass transition, which largely depends on the thermal history of the sample [11]. The temperature dependence of the dynamical behavior of amorphous sugar has been widely studied from a thermodynamic perspective [12].

On the other hand, the preparation dependence of the physical properties of amorphous sugar has also received much attention [13–15]. Surana et al. reported that together with aging, the preparation methods (i.e., freeze-drying, spray-drying, dehydration, and melt quenching) largely affect the glass transition temperature (T_g), the enthalpy relaxation behavior, the crystallization behavior, and the water sorption behavior of amorphous trehalose [13, 14]. They also mentioned that water vapor sorption could remove the structural history effects of the amorphous trehalose matrix formed by the dehydration method. More recently, Saxena et al. reported pronounced differences among the water sorption behaviors of amorphous trehalose prepared by different methods (annealing reversal, water sorption-desorption, and heating above T_g) [15]. These results strongly indicate that the properties of amorphous sugar materials are sensitive to water sorption and desorption. The importance of these findings in the pharmaceutical field cannot be overstated.

Specular X-ray reflectivity (XRR) is a unique and powerful analytical method (**Figure 2**) that is frequently applied to condensed soft-matter films, including glass transition studies of ultrathin polymer films [16–20], but which has not been applied in amorphous sugar studies. XRR can evaluate the layered structure of a material, such as the film thickness, electron density, surface roughness, and interfacial width [21, 22]. When the incident X-ray angle in the $2\theta/\omega$ scan exceeds the total reflection critical angle (θ_C), some of the X-rays penetrate the first layer, while the others reflect from the first-layer surface (**Figure 2a**). In turn, some of the penetrating X-rays enter the second layer, while the others reflect at that layer. The phase difference between the X-rays reflected from the two layers creates an interference pattern. Because the

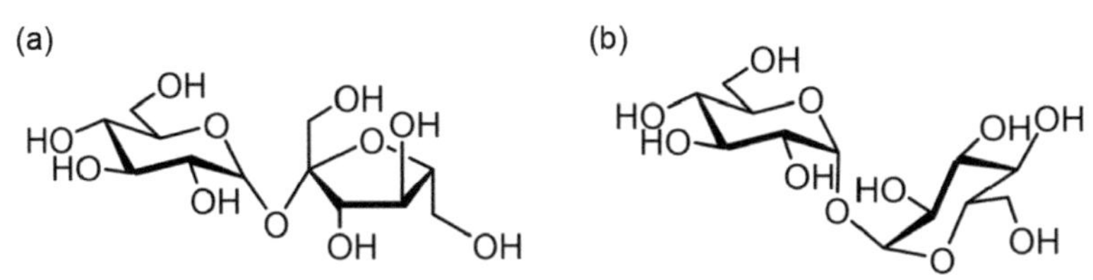

Figure 1. Molecular structures of two natural sugars: sucrose (Suc) (a) and trehalose (Tre) (b).

http://dx.doi.org/10.5772/66432

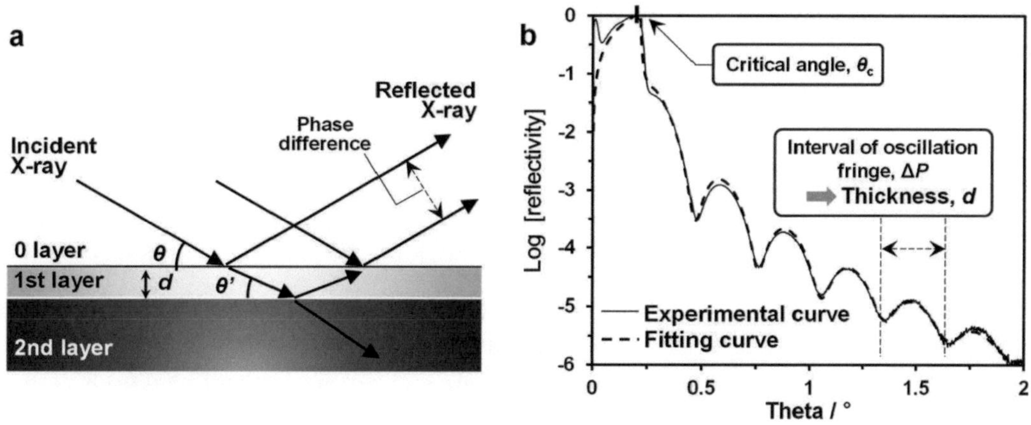

Figure 2. Schematic of XRR analysis (a) and a representative XRR profile, together with a fitting profile calculated by designated software (b).

observed X-ray scattering is the sum of the X-rays reflected by the electrons in each layer, the layers (including the substrate) are easily distinguished by their electron density differences. **Figure 2b** shows representative X-ray profiles of sugar thin film supported on silicon wafer substrate. When the X-ray reflectivity is plotted as a function of incident angle, oscillation fringes caused by X-ray interference appear in the profiles. Oscillation fringes are called "Kiessig fringes" after Kiessig, who first reported them in 1931 [23]. The oscillation period ΔP [radian] heavily depends on the film thickness, d; specifically, d is approximated by $\lambda/(2\Delta P)$, where λ is the X-ray wavelength [21], and is often calculated by fitting the X-ray profiles using an appropriate software [22]. In conventional X-ray optics with multilayer mirrors, this method is highly sensitive to the vertical length scale, which varies from 0.5 to 100 nm. Therefore, it has been extensively used to probe regions close to the interfaces of nano-films and is generally considered unsuitable for bulk material. Importantly, XRR is a nondestructive method, meaning that variations in the structural and morphological parameters of the film can be probed in the same sample. On account of these features, XRR analysis should also become a significant tool for investigating sugar nano-films, but this idea has been insufficiently studied.

The preparation of sugar film has been reported in several articles. Tre and Suc films have been prepared by drop casting [24] or spin coating [25–27] aqueous solutions of the sugar in atmospheric air or by vacuum deposition under reduced pressure [28]. In those reports, the films were thicker than 100 nm. On the other hand, Zhao et al. prepared glucose nano-films with thicknesses ranging from 20 to 60 nm [29]. They studied the adhesion and detachment behaviors of glassy, viscoelastic, and Newtonian liquid states of glucose and characterized the effect of the sugar glassy state on the surface deformations and flows. However, to our knowledge, Tre and Suc nano-films with thicknesses below 100 nm have not been fabricated, and detailed studies of their glass transitions and other properties have not been investigated. We prepared flat nano-films of Tre or Suc sugars by spin coating the aqueous sugar solutions onto silicon wafers and analyzed them by XRR. Large, fairly flat films are required, as the

XRR method has a very small angle of incidence. The effects of the vacuum operation and temperature on the spin-coated films were investigated, and the usefulness of XRR and some novel phenomenon related to the features of sugar glass are disclosed.

2. Experimental

2.1. Materials

Tre dihydrate, Suc, and ultrapure water were purchased from Wako Pure Chemical Co. Ltd. Si(100) substrate (thickness *ca.* 525 μm) was obtained from Electronics and Materials Corp. The Si(100) substrate was cut into a (2 × 2) cm^2 square and washed with ethanol prior to use.

2.2. Preparation of sugar nano-films

Tre and Suc were dissolved in ultrapure water by heating. Sugar nano-films were then prepared by spin coating the aqueous solutions on the Si(100) substrates at 4000 rpm for 45 s.

2.3. X-ray analyses on sugar nano-films

X-ray analyses of the sugar nano-films were performed in a multipurpose X-ray diffractometer (SmartLab, Rigaku Corp., Japan) equipped with a temperature control unit. Prior to measurements, the surface temperature calibration of the Si(100) substrate was checked by a conventional digital multimeter with a thermocouple wire. Specular XRR and X-ray diffraction (XRD) profiles were obtained by $2\theta/\omega$ scanning in the ranges θ = 0–2° and θ = 8–30°, respectively (Cu Kα, wavelength = 0.15418 Å) under air or vacuum conditions. The atmospheric pressure was reduced by a conventional oil hydraulic pump. The thickness and electron density of the films were determined by fitting the XRR profiles in the θ = 0–2° range using a numerical program developed in-house with Origin software. Our software adopted a recursive method based on dynamic scattering theory [30]. Grazing-incidence wide-angle X-ray diffraction (GI-XD) analysis was performed by 2θ scanning over the 2θ = 8–30° range at fixed incident angle (0.2°).

3. Results and discussion

3.1. Estimation of film thickness of spun-films

Figure 3 shows the XRR profile of the Si(100) substrate washed with ethanol. The X-ray reflection intensity was smoothly attenuated and no oscillation fringes were discernible at X-ray incidence angles above θ_C-Si (≈0.22°). This indicates that the surface layer of Si(100) was too thick or too thin for detectable interference between the media of different electron densities by XRR. In fact, the XRR profile fitting of the Si(100) substrate confirmed the absence of any layer thicker than 1 nm, meaning that the single crystal layer of the Si(100) substrate had been thoroughly cleaned. Sugar nano-films of different thicknesses were then prepared by spin coating aqueous sugar solutions of various concentrations onto the cleaned substrates.

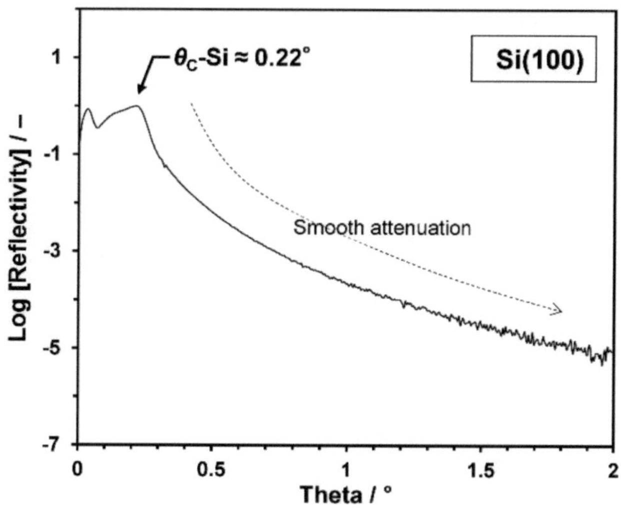

Figure 3. XRR profile of Si(100) substrate after washing with ethanol solvent. θ_c-Si is the critical angle of the Si wafer.

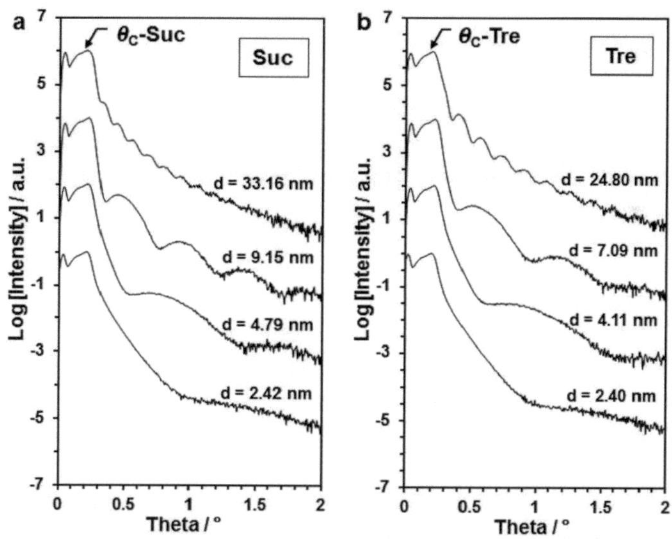

Figure 4. XRR profiles of Suc films (a) and Tre films (b) fabricated at a different thickness, *d*. Each profile represents a sugar film prepared by spin coating. The concentrations of the aqueous sugar solutions were 0.25, 0.50, 1.0, and 3.0 wt% from bottom to top. θ_C-Suc and θ_C-Tre are the critical angles of Suc and Tre films, respectively.

Figure 4 shows an XRR profile series of the spin-coated sugar films. The film thicknesses were calculated by fitting the XRR profiles. Along the series, the film thickness ranged from several to several dozens of nm, and clear oscillation fringes (Kiessig fringes) were observed in θ ranges above the critical incident angle θ_c-Suc and θ_c-Tre of Suc and Tre, respectively. The oscillation period ΔP appears to decrease with increasing concentration of the aqueous solution. As the thickness d is proportional to the reciprocal of ΔP [21], this result implies that

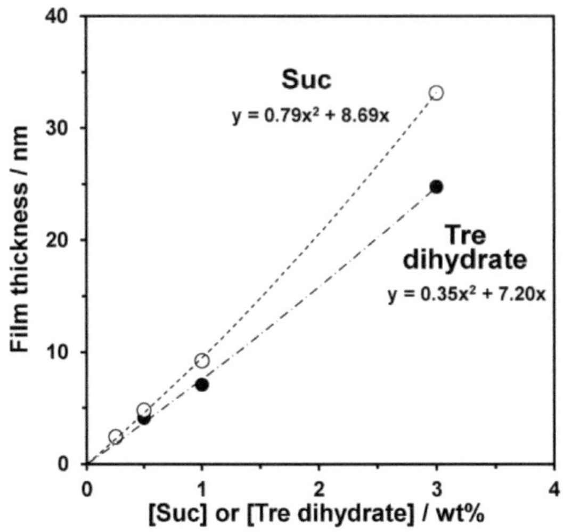

Figure 5. Thicknesses of sugar spun-films versus concentration of the corresponding aqueous sugar solutions used for spin coating. Open and closed circles denote the Suc and Tre sugar forms, respectively.

the higher the concentration of aqueous solution, the thicker the spin-coated film. The film thicknesses as functions of aqueous solution concentration are plotted in **Figure 5**. Each plot is well approximated by a quadratic equation.

3.2. Variation of film thickness of spun-films under reduced pressure

Water traces in the sugar matrix must be considered in studies of amorphous sugar. To confirm the existence of water, we analyzed the thicknesses of the Suc and Tre films in air and vacuum at room temperature and compared the pre-vacuum XRR profiles with those of the post-vacuum and subsequent vacuum release operations. **Figure 6** shows the XRR profiles analyzed (1) before and (2, 3) after the vacuum, together with the profiles (4) after vacuum release. From these results, we recognized that the vacuum operation increased ΔP from 1 to 2 and 3, indicating that the film thickness decreased after the vacuum operation was started. On the other hand, the vacuum release did not discernibly alter the thickness from the pre-release thickness, suggesting that the film thickness was reduced in the vacuum by water evaporation rather than the volume shrinkage of sugar molecules. To deny the hypothesis that the film thickness changed during a phase transition from amorphous to crystalline state, we analyzed the sugar nano-film by specular XRD and non-specular GI-XD. **Figure 7** compares the spectra of the films with those of the commercial powders of Suc and Tre dihydrate crystals. No diffraction peaks assignable to the crystalline structures were identified in the thin-film spectra, supporting that the sugar film did not crystallize during the experiment. Moreover, the XRR analysis detected the water desorption process in the sugar nano-films. After the vacuum operation, the thickness of the Suc spun-film (*ca.* 33.16 nm) had decreased by 5.45 nm, suggesting that water occupied 16.4 vol% in the sugar matrix (**Figure 6a**). In the Tre film of thickness *ca.* 24.7 nm, the vacuum operation removed a water content of approximately 4.4% (**Figure 6b**).

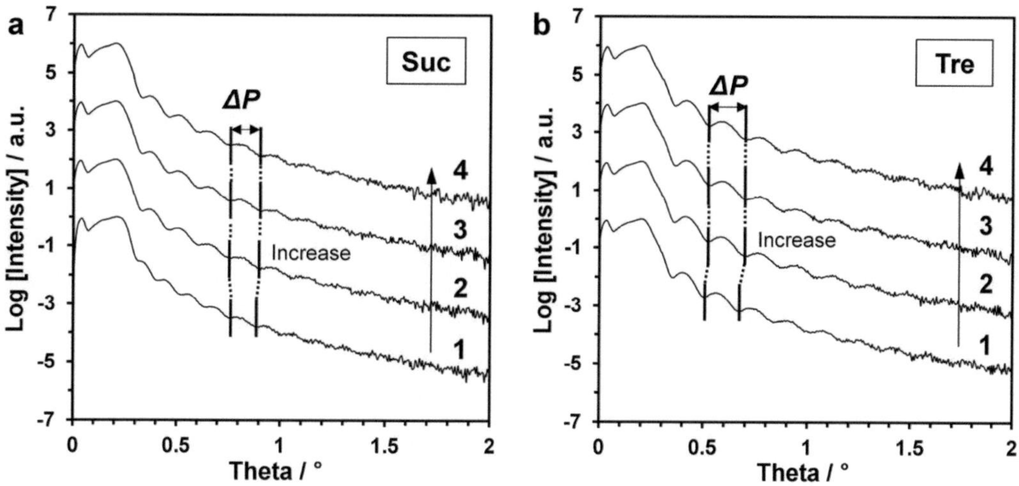

Figure 6. XRR profiles of sugar nano-films (1) before and (2, 3) after vacuum and (4) after vacuum release. Profiles 2 and 3 were obtained after vacuum operation for 15 min and 45 min, respectively. Profiles (1)–(4) represent Suc films with thicknesses of *ca.* 33.16, 27.86, 27.71, and 27.89 nm, respectively (a), and Tre films with thicknesses of *ca.* 24.79, 23.69, 23.45, and 23.52 nm, respectively (b).

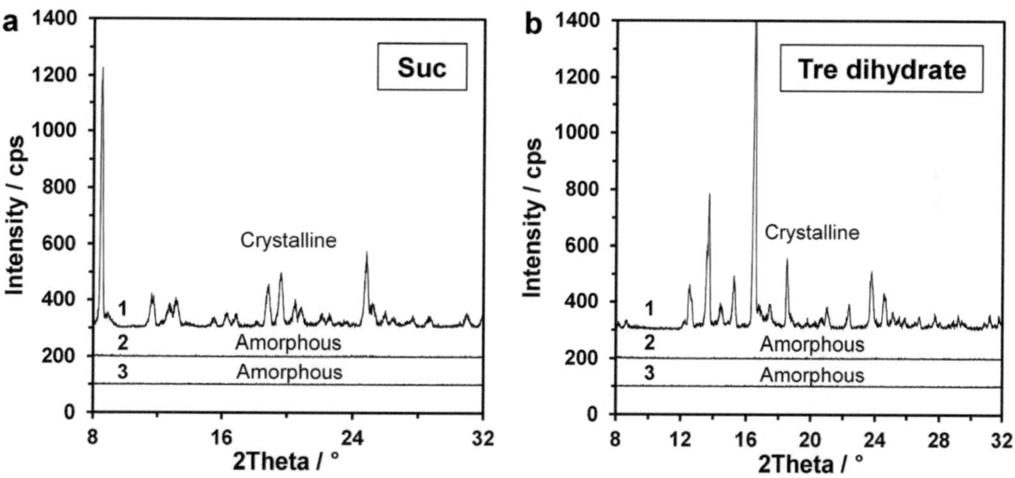

Figure 7. XRD profiles of (1) powder and (2) sugar nano-film and (3) GI-XRD profiles of sugar nano-films. The measured thicknesses of the Suc (a) and Tre (b) films were *ca.* 28 nm and 24 nm, respectively.

3.3. Variation of film thickness with temperature

3.3.1. Temperature-dependent behaviors of sugar nano-films

We next investigated the thickness of sugar nano-films fabricated at different temperatures. Experiments were conducted on Suc and Tre films with initial thicknesses of *ca.* 7.97 and 7.51 nm, respectively (**Figure 8**). Films of both sugars became thinner at higher temperatures (1 in **Figure 8**). The thicknesses of the Suc and Tre films decreased to *ca.* 7.84 nm at 90°C (−1.63%)

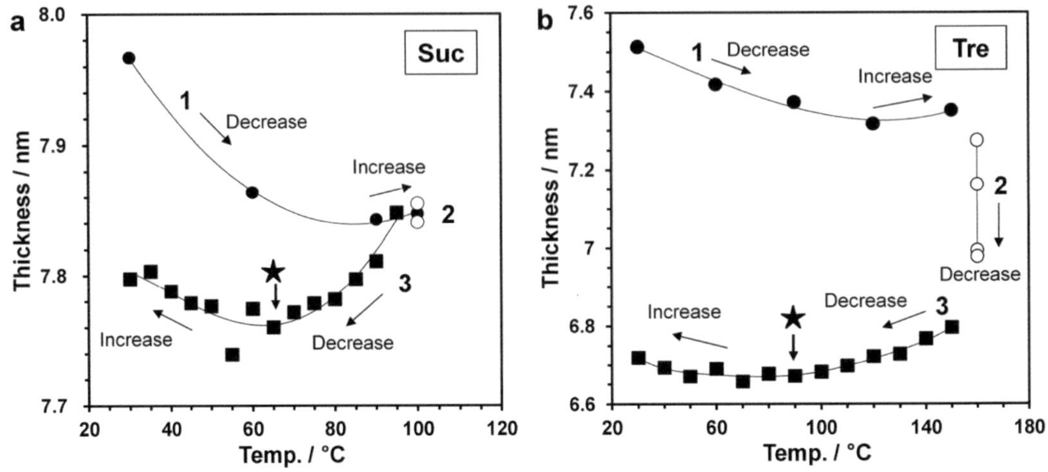

Figure 8. Temperature-dependent thickness changes in Suc and Tre films under vacuum conditions. Initially, the Suc (a) and Tre (b) films were *ca.* 8.0 nm and 7.5 nm thick, respectively. The Suc film was (1) heated from 30°C to 100°C, (2) held at 100°C for 80 min and then (3) cooled to 30°C. The Tre film was (1) heated from 30°C to 160°C, (2) held at 100°C for 150 min, and then (3) cooled to 30°C. The samples were heated and cooled at 5.0°C/min. The measurement time of each plot was 30 min.

and to *ca.* 7.32 nm at 120°C (−2.53%), respectively. No further decrease was observed after subsequent heating to 100°C for Suc film and to 160°C for Tre film (1 in **Figure 8**). Isothermal annealing at 100°C for 80 min did not alter the thickness of Suc film (2 of Suc in **Figure 8**), but isothermal annealing at 160°C reduced the thickness of the Tre film in a time-dependent manner (2 of Tre in **Figure 8**); this problem will be discussed in Section 3.3.3. Despite the different thickness responses of the sugar films at the isothermal annealing stage, the thicknesses of both sugars exhibited similar temperature-dependent behaviors during subsequent cooling (3 in **Figure 8**). Specifically, the film thicknesses decreased with cooling to a certain temperature and then increased with further cooling. The temperature at which the thickness stabilized (indicated by the filled star in the figure) was approximately 70°C for Suc film and 95°C for Tre film. As is well known, the volumes of conventional materials shrink with lowering temperature and its accompanying entropy decrease, but certain materials expand as the temperature reduces. The latter phenomenon, called negative thermal expansion (NTE) [17, 31], was apparently exhibited by the sugar nano-films during the cooling process.

The thermal expansion behavior of the Tre film depended on its thermal history (conditions are described in the caption of **Figure 9**). Similarly to **Figure 8**, the film thickness both decreased and increased during heating process (1 in **Figure 9**), and apparently decreased during isothermal annealing at 150°C (2 in **Figure 9a**). However, during subsequent cooling from 160° to 30°C, no NTE phenomenon was observed (3 in **Figure 9a**). In **Figure 9b** (Case II) the slopes of the normalized thickness versus temperature lines differ between above and below *ca.* 95°C. This temperature-dependent thickness behavior is probably attributable to the glass transition upon cooling. The glassy-to-viscous state transition of T_g is typically determined as the intersection of the two straight lines showing different linear expansivities [18]. In this way, we determined the T_g of trehalose nano-film as 95°C. We also compared the

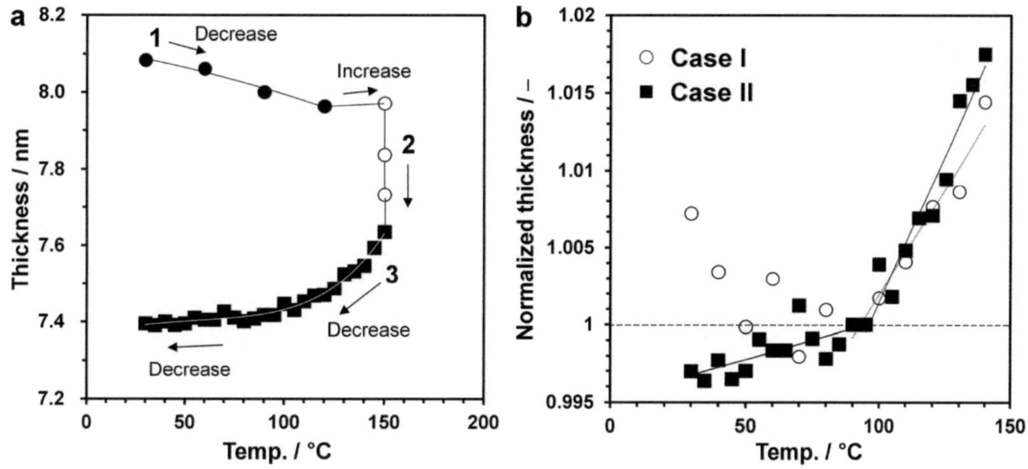

Figure 9. Temperature-dependent thickness changes in Tre film under vacuum conditions with different thermal history from that of **Figure 8b** (a). Initially, the Tre film was *ca.* 8.08 nm thick. The film was (1) heated from 30°C to 150°C, then (2) held at 150°C for 120 min, and (3) cooled to 30°C. The sample was heated and cooled at 10.0°C/min. Comparison of the thickness-temperature profiles in Cases I and II (b). Normalized thicknesses in Cases I and II result from the first cooling in **Figure 8b** and the first cooling in **Figure 9a**, respectively. Each film thickness was normalized by the film thickness at 90°C.

temperature-dependent thickness behaviors of Case I (after isothermal annealing at 160°C for 150 min) and Case II (after isothermal annealing at 150°C for 120 min). Notably, the slopes of the normalized thickness versus temperature lines were comparable above 95°C, despite the different thickness behaviors of Cases I and II below this temperature. From this result, the temperature indicated by the filled star in **Figure 8** was assumed as T_g. Above this temperature, the molecules can rapidly rearrange in the melt state, whereas below this temperature, their molecular mobility must be highly restricted.

3.3.2. Observation of reproducibility of NTE during heating and cooling cycle

NTE has received much attention as a tuning phenomenon in the overall thermal expansion of materials [31]. It has also been observed in ultrathin polymer films [16, 17, 20]. Mukherjee et al. reported NTE below the T_g in polystyrene (PS) films with thicknesses of several dozen nanometers, together with zero thermal expansion (ZTE) behavior [17]. They observed NTE and ZTE during both heating and cooling processes. To confirm the reproducibility of the NTE in the present study, we performed a reheating (second heating) operation. The thickness of the Tre film during reheating from 30°C to 140°C is depicted in **Figure 10a**. Although the second heating yielded slight deviations from the first cooling, the NTE behaviors were consistent in both processes, namely, the thickness decreased up to approximately 95°C and then increased with increasing temperature in the second heating process. Here, we estimated the area density of the electrons (average number of electrons per unit area of film) from the XRR profile and multiplied it by the film thickness. The results are plotted against temperature in **Figure 10b**. During the first cooling and second heating processes, the values were dispersed around the average value without large deviation, indicating that the NTE behavior does not

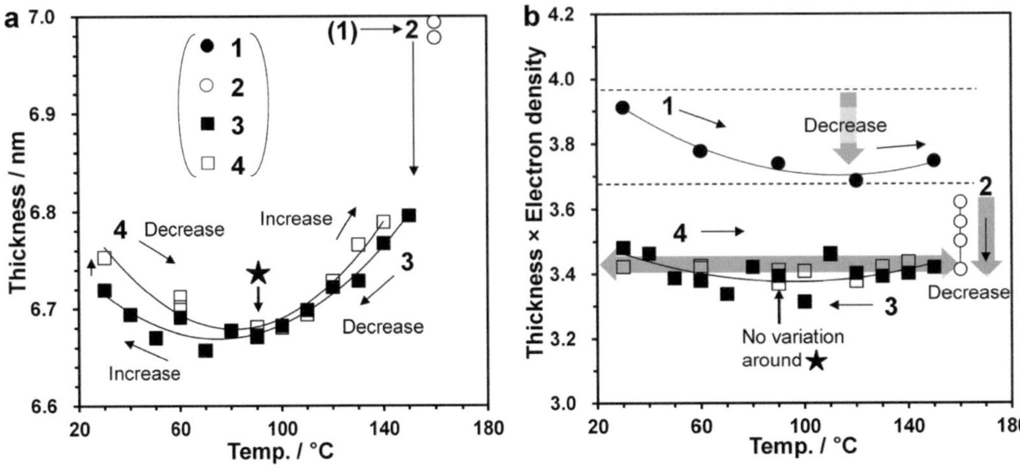

Figure 10. Thickness changes in Tre film during (3) the first cooling and (4) the second heating process (a) and the product of thickness and electron density during (1) the first heating, (2) isothermal annealing, (3) first cooling, and (4) second heating cycles (b). The heating and cooling were carried out at 5.0°C/min.

result from the absorption or desorption of trace amounts of water. Also, the temperature-dependent variation in film thickness was confirmed to occur perpendicularly to the substrate, indicating that the density of the sugar matrix varies with temperature because the film thickness varies with temperature. The film thickness should therefore increase with decreasing density.

In **Figures 7–9**, an apparent hysteresis appears between the thicknesses during the first heating and first cooling and during the first and second heating processes. During the first heating process, the thickness (and its product with the electron density) monotonically decreases with increasing temperature up to 90°C and 120°C for Suc and Tre film, respectively. This effect might be attributable to the evaporation of trace amounts of water during the heating. The filling of the voids left by the departing water molecules might shrink the sugar matrix. Above 90°C for Suc film and 120°C for Tre film, the thickness and electron density product became invariant, suggesting that the water had wholly evaporated. Thus, NTE was observed during the first heating process, but apparently the first cooling process was governed by different mechanisms.

3.3.3. Sublimation problem

The dominant factor governing the inclusion or exclusion of NTE is unclear, but a likely scenario is rearrangement of the sugar matrix in the melt state, as the isothermal conditions are different in Cases I and II. In fact, the largely reduced thickness of the Tre film after isothermal annealing at 160°C was unexpected (see process 2 in **Figure 8b**). The product of thickness and electron density in Tre film exhibited a similar trend (2 in **Figure 10b**). **Figure 11** plots the time-dependent thickness of the Tre film under isothermal annealing at 170°C. Prior to annealing, the film was gradually heated to 150°C. During the heating process, the decrease and subsequent increase in thickness mimicked that of stage (1) in **Figure 8** (**Figure 11b**). In

http://dx.doi.org/10.5772/66432

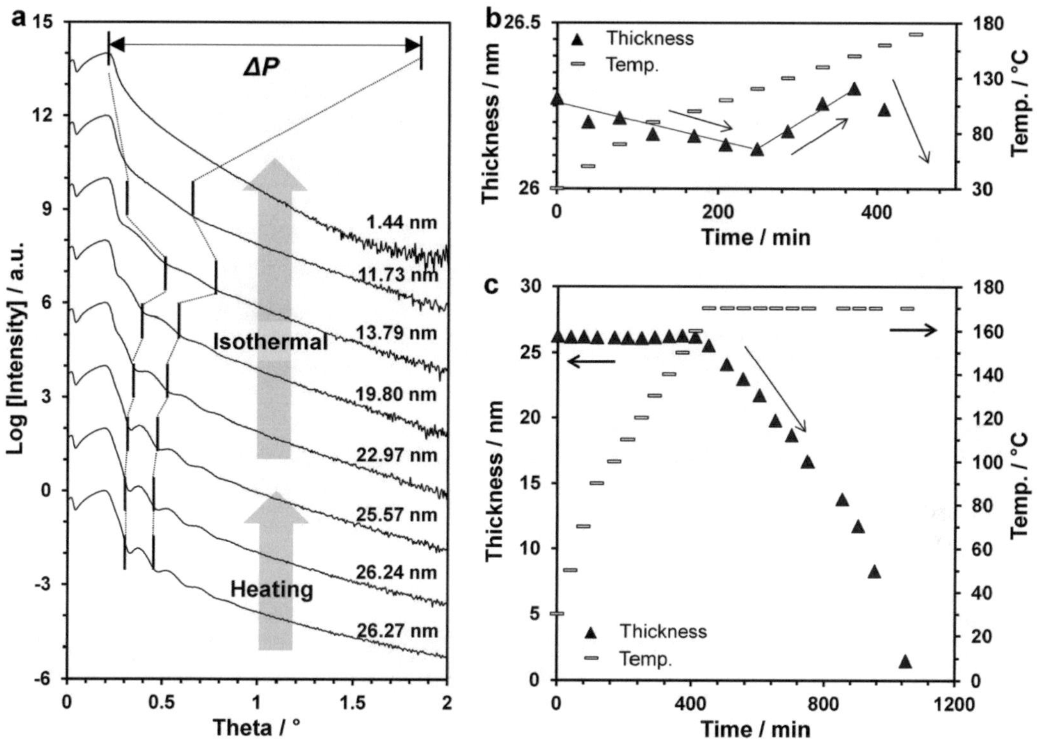

Figure 11. XRR profiles (a) and changing thickness of Tre film under the vacuum condition during heating up to 160°C (b) and during isothermal annealing at 170°C (c).

addition, the thickness began decreasing at 160°C. Subsequently, under isothermal annealing at 170°C, the film thickness continuously decreased, reaching 0 after *ca.* 10 h (**Figure 11c**), as clearly evidenced by the broadening of ΔP in the XRR profiles (**Figure 11a**). This result strongly supports sublimation of the Tre film under high-temperature low-pressure conditions. Because Tre is a small molecule with a molecular weight of 342.3, it can potentially sublimate under these conditions. During sublimation, the molecules at the air-sugar interface are easily rearranged, which would largely influence the whole range of nano-films. Therefore, the isothermal annealing conditions can affect the states of the nano-films and determine the occurrence or absence of NTE. For a more detailed discussion, we should accurately tune the reduced pressure condition to control or prevent the sugar sublimation.

4. Conclusions

In this chapter, we demonstrated the temperature-dependent thickness behaviors of sugar nano-films formed on Si(100) substrates of area (2 × 2) cm^2. Homogeneous flat films suitable for precise XRR analysis were fabricated by the conventional spin-coating method. The complex behaviors of the shrinking and expanding sugar nano-films during heating and cooling were successfully observed and are schematically summarized in **Figure 12**.

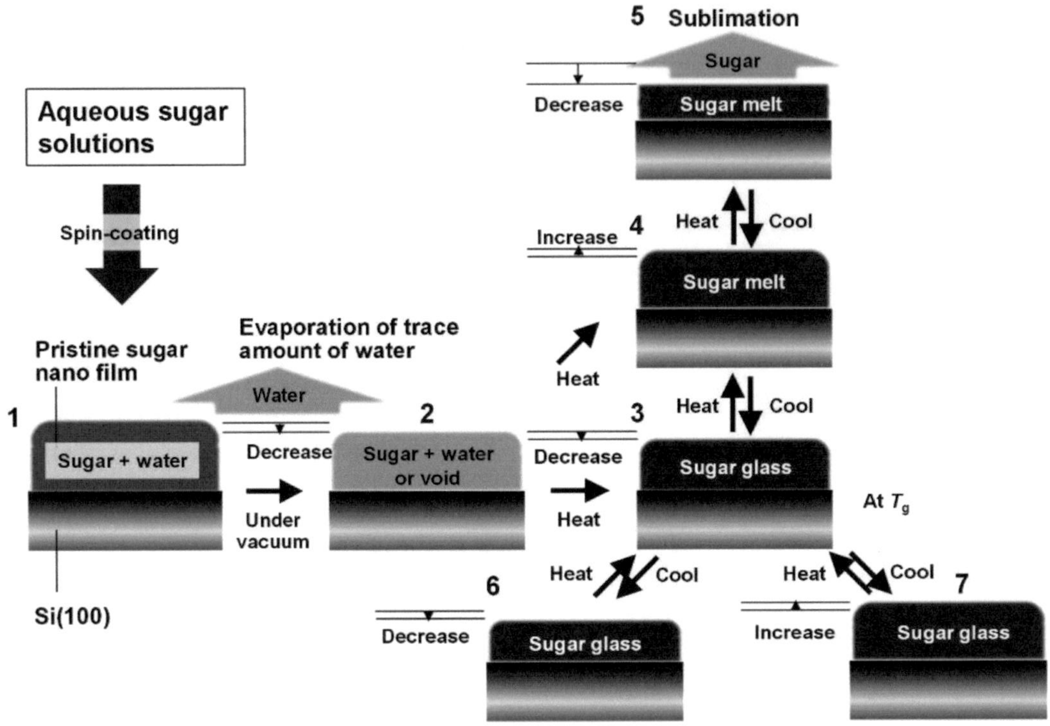

Figure 12. Schematic representation of the behavior of sugar nano-films under reduced pressure.

Once the aqueous sugar solutions have been spin-coated on the Si substrates, residual water remains in the nano-films (1 in **Figure 12**). Such water can be mostly or partly removed by evaporation under reduced pressure even at room temperature (1–2 in **Figure 12**). Evaporation manifests as decreased film thickness. As the T_gs of Suc and Tre are much higher than room temperature, an anhydrous sugar matrix system must exist in the glassy state at room temperature. Otherwise, the system is a sugar matrix adsorbing trace amounts of water. The film thickness was generally decreased by heating operations (2–3 or 4 or the intermediate state of 3 and 4 in **Figure 12**). This decrease is ascribed to the evaporation of remaining water in the sugar matrix. The evaporated water leaves voids that are then filled by contraction of the sugar matrix during the subsequent glass formation. The glassification must prevent the short-time rearrangement of molecules at temperatures below T_g, but as the sugar molecules gain thermal energy when heated above T_g, they can readily rearrange in the matrix. However, under reduced pressure, the relatively low-molecular-weight natural sugars inevitably sublimate, as they lack a nonvolatile property (5 in **Figure 12**). This situation never arises in polymeric film. Beside sublimating themselves in the matrix, the sugar molecules redistribute at the vacuum-sugar interface, which should significantly affect the overall nano-film locating around the interface. As the temperature decreases, sublimation becomes less important, and the thickness might decrease through shrinkage of the sugar matrix under thermodynamic processes. Below T_g, the sugar films showed two behaviors, a normal reduction in film thickness with smaller expansibilities than obtained between 3 and 4 in **Figure 12** (3–6 in **Figure 12**) and an unusual increase in film thickness governed by NTE (from 3 to 7 in **Figure 12**). These processes appear not to be determined by the initial film thickness; rather, they depend on the thermal history, for reasons which are currently unknown. To understand the effect of molecular rearrangement at the vacuum-sugar interface, we should study anhydrous sugar nano-films under precisely defined vacuum conditions.

As shown in this chapter, the XRR methodology provides profound insights into the adsorption and desorption properties of amorphous sugars, the rearrangement of sugar molecules at the sugar-air interface and the glass transition. To acquire these insights, we observed how the film thickness depends on water content and temperature. By understanding the sugar-vacuum or sugar-solid interface, we might also capture the structural changes of sugar matrixes under freezing, freeze-drying, and spray-drying operations. In all of these processes, molecular transfer such as water adsorption and desorption starts at the interface.

Acknowledgement

This work was financially supported by the Amano Institute of Technology, AIST, Japan, by the MEXT-Supported Program for the Strategic Research Foundation at Private Universities (S1201027) 2012–2016, and by the Japan Society for the Promotion of Science (Grant-in-Aid for Scientific Research (C) 24560033).

Author details

Shigesaburo Ogawa[1] and Isao Takahashi[2*]

*Address all correspondence to: suikyo@kwansei.ac.jp

1 Department of Materials and Life Science, Faculty of Science and Technology, Seikei University, Japan

2 Department of Physics, School of Science and Technology, Kwansei Gakuin University, Japan

References

[1] Kauzmann W: The nature of the glassy state and the behavior of liquids at low temperatures. Chemical Reviews. 1948;43:219–256. DOI: 10.1021/cr60135a002

[2] MacKenzie AP: Non-equilibrium freezing behavior of aqueous systems. Philosophical Transactions of the Royal Society B: Biological Sciences. 1977;278:167–189. DOI: 10.1098/rstb.1977.0036

[3] Slade L, Levine H: Non-equilibrium behavior of small carbohydrate-water systems. Pure and Applied Chemistry. 1988;60:1841–1864. DOI: 10.1351/pac198860121841

[4] Carpenter JF, Crowe LM, Crowe JH: Stabilization of phosphofructokinase with sugars during freeze-drying: characterization of enhanced protection in the presence of divalent cations. Biochimica et Biophysica Acta. 1987;923:109–115. DOI: 10.1016/0304-4165(87)90133-4

[5] Slade L, Levine H, Reid DS: Beyond water activity: recent advances based on an alternative approach to the assessment of food quality and safety. Critical Reviews in Food Science & Nutrition. 1991;30:115–360. DOI: 10.1007/978-1-4899-0664-9_3

[6] Fox KC: Biopreservation. Putting proteins under glass. Science. 1995;267:1922–1923. DOI: 10.1126/science.7701317

[7] Dave H, Gao F, Lee JH, Liberatore M, Ho CC, Co CC: Self-assembly in sugar–oil complex glasses. Nature Materials. 2007;6:287–290. DOI: 10.1038/nmat1864

[8] Santivarangkna C, Higl B, Foerst P: Protection mechanisms of sugars during different stages of preparation process of dried lactic acid starter cultures. Food Microbiology. 2008;25:429–441. DOI: 10.1016/j.fm.2007.12.004

[9] Giri J, Li W-J, Tuan RS, Cicerone MT: Stabilization of proteins by nanoencapsulation in sugar-glass for tissue engineering and drug delivery applications. Advanced Materials. 2011;23;4861–4867. DOI: 10.1002/adma.201102267

[10] Mittal A, Schulze K, Ebensen T, Weissmann S, Hansen S, Guzmán CA, Lehr CM: Inverse micellar sugar glass (IMSG) nanoparticles for transfollicular vaccination. Journal of Controlled Release. 2015;206:140–152. DOI: 10.1016/j.jconrel.2015.03.017

[11] Simatos D, Blond G, Roudaut G, Champion D, Perez J, Faivre AL: Influence of heating and cooling rates on the glass transition temperature and the fragility parameter of sorbitol and fructose as measured by DSC. Journal of Thermal Analysis. 1996;47:1419–1436. DOI: 10.1007/BF01992837

[12] Angell CA: Liquid fragility and the glass transition in water and aqueous solutions. Chemical Reviews. 2002;102:2627–2650. DOI: 10.1021/cr000689q

[13] Surana R, Pyne A, Suryanarayanan R: Effect of preparation method on physical properties of amorphous trehalose. Pharmaceutical Research. 2004;21:1167–1176.

[14] Surana R, Pyne A, Suryanarayanan R: Effect of aging on the physical properties of amorphous trehalose. Pharmaceutical Research. 2004;21:867–874. DOI: 10.1023/B:PHAM.0000026441.77567.75

[15] Saxena A, Jean YC, Suryanarayanan R: Annealing effect reversal by water sorption–desorption and heating above the glass transition temperature-comparison of properties. Molecular Pharmaceutics. 2013;10:3005–3012. DOI: 10.1021/mp400099r

[16] Orts WJ, van Zanten JH, Wu W-L, Satija SK: Observation of temperature dependent thickness in ultrathin polystyrene films on silicon. Physical Review Letters. 1993; 71:867–870. DOI: http://dx.doi.org/10.1103/PhysRevLett.71.86

[17] Mukherjee M, Bhattacharya M, Sanyal MK, Geue T, Grenzer J, Pietsch U: Reversible negative thermal expansion of polymer films. Physical Review E. 2002;66:061801. DOI: 10.1103/PhysRevE.66.061801

[18] Takahashi I, Yang C: Broadening, no broadening and narrowing of glass transition of supported polystyrene ultrathin films emerging under ultraslow temperature variations. Polymer Journal. 2011;43:390–397. DOI: 10.1038/pj.2010.145

[19] Sun X, Guo L, Sato H, Ozaki Y, Yan S, Takahashi I: A study on the crystallization behavior of poly(β-hydroxybutyrate) thin films on Si wafers. Polymer. 2011;52:3865–3870. DOI: http://dx.doi.org/10.1016/j.polymer.2011.06.024

[20] Gin P, Jiang N, Liang C, Taniguchi T, Akgun B, Satija SK, Endoh MK, Koga T: Revealed architectures of adsorbed polymer chains at solid-polymer melt interfaces. Physical Review Letters. 2012;109:265501. DOI: 10.1103/PhysRevLett.109.265501

[21] Tolan M: Springer Tracts in Modern Physics. X-ray Scattering from Soft-Matter Thin Films. Materials Science and Basic Research. Springer: Berlin, vol. 148; 1999. 197 p. DOI: 10.1007/BFb0112834

[22] Yasaka M: X-ray thin-film measurement techniques V. X-ray reflectivity measurement. The Rigaku Journal. 2010;26:1–9.

[23] Kiessig H: Investigation to the total X-ray reflection. Annalen der Physik. 1931;10:715–768. DOI: 10.1002/andp.19314020607

[24] Wright WW, Guffanti GT, Vanderkooi JM. Proteins in sugar films and in glycerol/water as examined by infrared spectroscopy and by the fluorescence and phosphorescence of tryptophan. Biophysical Journal. 2003;85:1980–1995. DOI: http://dx.doi.org/10.1016/S0006-3495(03)74626-8

[25] Cheng J, Wingrad N: Depth profiling of peptide films with TOF-SIMS and a C60 probe. Analytical Chemistry. 2005;77:3651–3659. DOI: 10.1021/ac048131w

[26] Zhu Z, Nachimuthu P, Lea AS: Molecular depth profiling of sucrose films: a comparative study of C_{60}^{n+} ions and traditional Cs^+ and O_2^+ ions. Analytical Chemistry. 2009;81:8272–8279. DOI: 10.1021/ac900553z

[27] He X, Fowler A, Menze M, Hand S, Toner M: Desiccation kinetics and biothermodynamics of glass forming trehalose solutions in thin films. Annals of Biomedical Engineering. 2008;36:1428–1439. DOI: 10.1007/s10439-008-9518-8

[28] Predoi D: Physico-chemical studies of sucrose thin films. Digest Journal of Nanomaterials and Biostructures. 2010;5:373–377.

[29] Zhao B, Zeng H, Tian Y, Israelachvili J: Adhesion and detachment mechanisms of sugar surfaces from the solid (glassy) to liquid (viscous) states. Proceedings of the National Academy of Sciences. 2006;103:19624–19629. DOI: 10.1073/pnas.060952910

[30] Parratt LG: Surface studies of solids by total reflection of X-rays. Physical Review. 1954;95:359–369. DOI: http://dx.doi.org/10.1103/PhysRev.95.359

[31] Takenaka K: Negative thermal expansion materials: technological key for control of thermal expansion. Scientific and Technology of Advanced Materials. 2012;13:013001. DOI: 10.1088/1468-6996/13/1/013001

Chapter 7

Use of Ionic Liquids for the Treatment of Biomass Materials and Biofuel Production

El-Sayed R.E. Hassan and Fabrice Mutelet

Additional information is available at the end of the chapter

http://dx.doi.org/10.5772/67026

Abstract

Biomass, as fuel source, is renewable, environmental friendly and abundant in nature. It is of great interest to produce green energy and bio-products from lignocellulose. The replacement of conventional organic solvents by a new generation of solvents that are less toxic, less flammable and less polluting is a major challenge for the chemical industry. The aim of this work is to study the solubility of biomass-based materials in ionic liquids in order to overcome the lack of experimental data on phase equilibria of {carbohydrate-ILs} mixtures. Solubility data were successfully correlated using NRTL and UNIQUAC thermodynamic models. The fundamental natures of the interaction between carbohydrates and ILs were investigated using ab initio calculations. The pretreatment of miscanthus with ILs resulted in the regeneration of amorphous, porous cellulose almost free of lignin, which is suitable for enzymatic hydrolysis and fermentation processes. A successful ethanol production was obtained with an overall ethanol yield reached up to 150 g ethanol kg^{-1} miscanthus.

Keywords: ionic liquids, biomass, biofuel, miscanthus, ab initio calculations

1. Introduction

Nowadays, governments and researchers are directing their attention toward the development of advanced biofuels. Lignocellulosic biomass is an alternative raw material for biofuel production [1, 2]. Lignocellulosic biomass materials are woody and straw residues from agriculture and forestry, nonfood crops, organic fraction of urban waste and algae-based feedstock. Lignocellulosic biomass consists mainly of lignin, hemicellulose and cellulose [3]. The direct production of sugars from biomass is expensive and inefficient if compared with starch-based feedstocks due to its complicated structure. Therefore, the first step of converting biomass into biofuel is the pretreatment of biomass to break down the lignin-carbohydrate complex structure and to decrease the crystallinity of the extracted cellulose. The second step is the hydrolysis of extracted cellulose to fermentable sugars such as glucose and the final step is the fermentation of the sugars into liquid fuels [4, 5].

The pretreatment of biomass has been achieved using different technologies in order to extract cellulose. These technologies include ammonia fiber expansion, sodium chlorite and ethanol organosolv process [6]. These technologies showed acceptable results concerning the cellulose extraction. Nevertheless, the recovery of the lignin and hemicelluloses should be taken into consideration in order to optimize the production of bioethanol from biomass.

The replacement of conventional organic solvents by a new generation of solvents less toxic, less flammable and less polluting is a major challenge for the chemical industry. Ionic liquids have been widely promoted as interesting substitutes for traditional solvents. Ionic liquids are salts in the liquid state having melting point below 373 K [7]. They have unique physicochemical properties such as high thermal stability nonflammability, electroconductivity and negligible vapor pressure. Therefore, they are considered as more environmentally friendly than their volatile, toxic and organic counterparts. Ionic liquids, as green solvents, have been widely applied in important fields such as catalysis, separation and nanotechnology [8].

The solubility data of biomass materials in ILs are of great importance for chemical or bioprocesses. Indeed, a small number of studies have been published on the solubility of biomass materials in ionic liquids. Swatloski et al. found that 1-butyl-3-methylimidazolium chloride (BMIMCl) is able to dissolve up to 25 wt%. of cellulose [9]. Sheldon et al. connected ILs with carbohydrates [10]. Carneiro et al. studied the solubility of monosaccharides in ILs and correlated these data using the NRTL and UNIQUAC thermodynamic models [11].

Currently, ab initio calculations play an important role in understanding the special nature of ILs and their interactions with dissolved components or interfaces [12]. Hydrogen bonding in ILs plays an important role in cation-anion and solvent-solute interactions. This phenomenon could be revealed by both experimental and theoretical investigations. In fact, the hydrogen bonding of ILs is considered to be a factor of great importance to design ILs as potential solvents for cellulose. Xu et al. [12] showed that both chloride anions and imidazolium cations of the IL interact with the cellulose via hydrogen bonds. The results emphasize that the chloride anions play a critically important role and the imidazolium cations present a remarkable contribution in the cellulose dissolution.

Miscanthus, as a biomass feedstock, is an interesting raw material for biofuel production, as it has a carbohydrate content up to 75% [6]. Indeed, there is a lack of data concerning the ionic liquid pretreatment, enzymatic hydrolysis and fermentation of miscanthus. Shill et al. reported the solubility of miscanthus and the extraction of cellulose using EMIMAOC ionic liquid [13]. Padmanabhan et al. found that acetate, chloride and phosphate imidazolium-based ionic liquids are able to dissolve miscanthus up to 5% [14].

Enzymatic hydrolysis process has many advantages such as the very mild conditions (pH = 4.8 and temperature 323 K) which give high yields. The maintenance costs are low compared to alkaline and acid hydrolysis due to no corrosion problems. Hydrolysis without preceding pretreatment yields typically lower than 20%, whereas it yields after pretreatment often exceed

http://dx.doi.org/10.5772/67026

90%. Yeast is currently the most popular method for converting cellulosic sugars into ethanol. The most common type of yeast used is *Saccharomyces cerevisiae*, also known as Brewer's yeast or Baker's yeast. *S. cerevisiae* has a relatively high tolerance to ethanol and inhibitor compounds. Furthermore, this yeast gives high ethanol yields from glucose [3].

This chapter presents an environmentally friendly method for extracting cellulose from biomass using imidazolium-based ionic liquids and also provides a new approach for utilizing biomass resources. This chapter is mainly divided into three parts: The aim of the first part is to overcome the lack of experimental data on phase equilibria of biomass carbohydrates in ionic liquids. The solubility of biomass materials in ionic liquids was measured within a temperature range from 283 to 383 K. Solubility data were correlated successfully with local composition thermodynamic models such as NRTL and UNIQUAC. The solubility of biomass materials in binary mixtures {ethanol + ionic liquids} is studied in order to evaluate the possible use of the antisolvent method for the extraction of carbohydrates from ionic liquids. The second part is devoted to investigate the interaction between carbohydrates, glucose and cellulose building unit and ionic liquids using ab initio quantum chemical methods. The third part aims to study the solubility of miscanthus, as a promising biomass feedstock, in ILs and cellulose extraction. The parameters affecting the extraction process are the ionic liquid structure, miscanthus size, miscanthus mass fraction, temperature and time. A Box-Benhken design expert was applied to evaluate the best conditions for the extraction process. Finally, the produced amorphous cellulose is subjected to hydrolysis and fermentation for biofuel production.

2. Study of the behavior of systems containing (carbohydrate-ILs)

The aim of this part is to overcome the lack of experimental data on phase equilibria of biomass carbohydrates in ionic liquids. The solubility of glucose, fructose, sucrose and lactose in five ionic liquids: 1-ethanol-3-methylimidazolium chloride, 1-butyl-1-methylpyrrolidinium chloride, 1,3-dimethyl-imidazolium methyl phosphonate, BMIMCl and EMIMSCN, was measured within a temperature range from 283 to 383 K (**Table 1**).

2.1. Solubility of carbohydrates in pure ionic liquids

Solid-liquid equilibria (SLE) of binary systems {IL + sugar} were carried out in a large range of temperatures 280–390 K and compositions up to 60% of sugar, (**Figure 1**). It is clearly obvious that the solubility of carbohydrates increases in the following order lactose < sucrose < glucose < fructose for ILs used in this study. The solubility increases at high temperature in the following order EMIMSCN < EtOHMIMCl ≤ BMIMCl < DMIMMPh. The Kamlet-Taft solvatochromic parameters are the most comprehensive and frequently used quantitative measure of solvent properties. Studies on the dissolution of cellulose in BMIMCl indicate that the anion of the IL acts as a hydrogen bond acceptor which interacts with the hydroxyl groups of the cellulose [9]. It is found that the phosphonate and chloride imidazolium-based ionic liquids displayed higher β and π^* values if compared to other ionic liquids. It was found that the solubility of sugars is related to the hydrogen bond basicity and the polarizability of the ionic liquids.

Physical properties	BMIMCl	EtOHMIMCl	BMPyCl	EMIMSCN	DMIMMPh
Color and shape at 25°C	White powder	Yellow solid	White powder	Red liquid	Colorless liquid
Molecular formula	$C_8H_{15}ClN_2$	$C_6H_{11}ClON_2$	$C_9H_{20}ClN$	$C_7H_{11}N_3S$	$C_6H_{13}N_2PO_3$
Purity (%)	98	98	98	95	98
Molecular weight	174.7	162.61	177.71	162.25	192.15
Melting point (°C)	65	80	114	−6	−20
Density (g/cm^3)	1.0528 at 80°C	n.d	n.d	1.114 at 25°C	1.18 at 25°C
Viscosity (cP)	142 at 80°C	n.d	n.d	20 at 25°C	50.3 at 25°C
Decomposition temp. (°C)	250	>200	210	>200	>200
Water content (ppm)	<1000	<1000	<1000	<1000	450
Solubility in water	Miscible	Miscible	Miscible	Miscible	Miscible
Solubility in ethanol	Miscible	Miscible	Miscible	Miscible	Miscible

Table 1. Physical properties of studied ionic liquids.

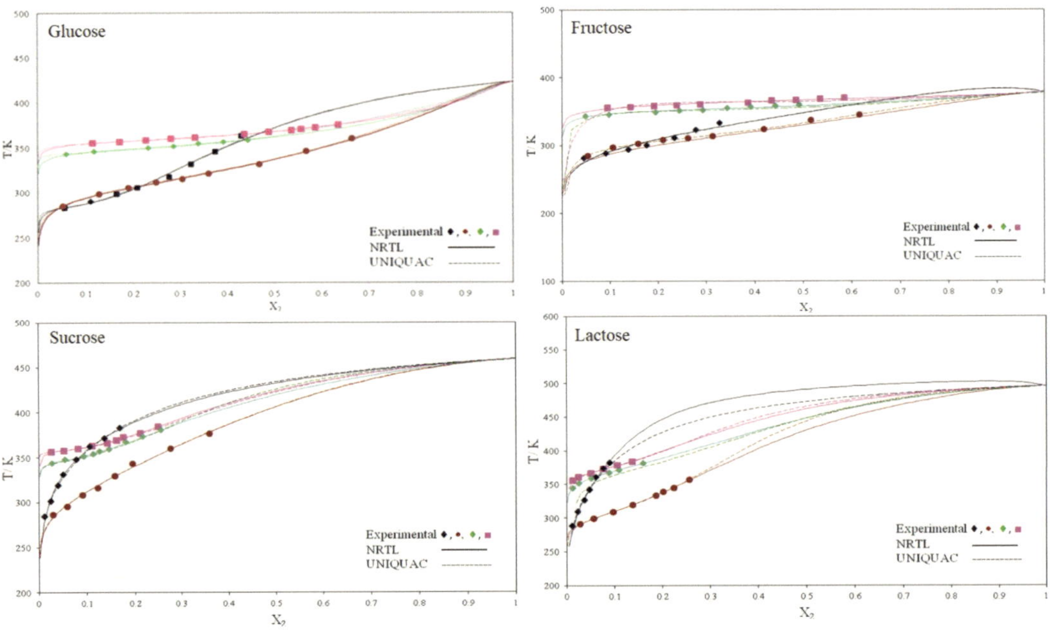

Figure 1. Plot of the experimental and calculated SLE of {IL + sugar} binary systems: ♦ EMIMSCN; ● DMIMMPh; ♦ BMIMCl; ▪ EtOHMIMCl. The solid lines have been calculated using NRTL and UNIQUAC models, where X_2 is the sugar mole fraction.

Figure 1 shows that solubility of carbohydrates is a complex process in ILs. Indeed, glucose is more soluble in EMIMSCN than in DMIMMPh up to 310 K. At higher temperatures, the opposite effect is observed. The main key factors of solubility of biomass components in ILs are physicochemical properties such as viscosity, basicity and polarity. Experimental data on the solubility of glucose in thiocyanate-based ILs proved that the solubility of glucose increases with a decrease of the alkyl chain length [15, 16]. **Figure 1** points out the importance of the influence of functional group in ILs [17]. As an example, the presence of a hydroxyl group in ILs may lead to hydrogen bond formation with the anion or the cation and then modify the behavior of the solvent with carbohydrates.

Experimental data found in the literature on the solubility of fructose and sucrose in ILs proved that their solubility is strongly affected by the alkyl chain length attached on the IL but also the basicity and the polarity. The solubility of sucrose increases with increasing the hydrogen bond basicity and polarity of the anion of the ionic liquid in the following order: $C(CN)_3 < BF_4 < CF_3SO_3 < CH_3SO_3 < SCN < HSO_4 < Cl < CH_3HPO_3$ [10, 11, 15, 18]. Moreover, the solubility data of sucrose in dialkylimidazolium thiocyanate measured in this work, found in the literature [15] and confirmed that the solubility of carbohydrates decreases with an increase of the alkyl chain length grafted in the cation.

2.2. Estimation of the solubility of carbohydrates in ILs

It is well established that solid-liquid equilibria of systems containing ILs can be well represented with thermodynamic models such as nonrandom two-liquid equation (NRTL) [19] or the UNIversal QUAsiChemical (UNIQUAC) theory [20]. The solubility of the carbohydrates in the ILs was determined from an expression based on the symmetric convention for the calculation of the activity coefficients, that is, the pure liquid at the solution temperature as a standard state for the carbohydrate, together with their fusion enthalpy, $\Delta_{\mathrm{fus}}H$ and melting temperature, T_{fus}. For a general case, this equation can be derived through an idealized thermodynamic cycle between the solid and liquid carbohydrate phase states, under this assumption: The solvent does not appear in the solid phase. Thus, the resulting expression to the calculation of the solubilities is [21] as follows:

$$\ln(x_2\gamma_2) = \frac{-\Delta_{fus}H_2}{R}\cdot\left(\frac{1}{T^{SLE}}-\frac{1}{T_{fus,2}}\right) \tag{1}$$

where $\Delta_{\mathrm{fus}}H_2$ and T_{fus2} denote melting enthalpy and temperature of the sugar and x_2 stands for the solubility of the sugar at the saturated temperature T^{SLE}.

2.2.1. NRTL model

For the NRTL model, the activity coefficient γ_i, for any component i of the ternary system, is given by:

$$\ln\gamma_i = \frac{\sum_{j=1}^{m}\tau_{ji}G_{ji}x_j}{\sum_{l=1}^{m}G_{li}x_l} + \sum_{j=1}^{m}\frac{x_jG_{ij}}{\sum_{l=1}^{m}G_{lj}x_l}\left(\tau_{ij}-\frac{\sum_{r=1}^{m}x_r\tau_{rj}G_{rj}}{\sum_{l=1}^{m}G_{lj}x_l}\right) \tag{2}$$

with $G_{ji} = \exp(-\alpha_{ji}\tau_{ji})$, $\tau_{ji} = \frac{g_{ji}-g_{ii}}{RT} = \frac{\Delta g_{ji}}{RT}$ and $\alpha_{ji} = \alpha_{ij} = \alpha$

where g is an energy parameter characterizing the interaction of species i and j, x_i is the mole fraction of component i. In this work, the nonrandomness parameter α was set equal to 0.2 according to the literature [22].

2.2.2. *UNIQUAC model*

For the UNIQUAC model, the activity coefficient γ_i, for any component i of the binary system, is given by:

$$\ln\gamma_i = \ln\frac{\Phi_i}{x_i} + \frac{z}{2}q_i\ln\frac{\theta_i}{\Phi_i} + l_i - \frac{\Phi_i}{x_i}\sum_{j=1}^{m} x_j l_j - q_i\ln(\theta_j\tau_{ji}) + q_i - q_i\sum_{j=1}^{m}\frac{\theta_j\tau_{ji}}{\sum_{k=1}^{m}\theta_k\tau_{kj}} \quad (3)$$

where $\Phi_i = \frac{r_i x_i}{\sum_{j=1}^{m} r_j x_j}$, $\theta_i = \frac{q_i x_i}{\sum_{j=1}^{m} q_j x_j}$, $l_j = \frac{z}{2}(r_j - q_j) - (r_j - 1)$ and $\tau_{ji} = \exp\left(\frac{-\Delta u_{ij}}{RT}\right)$

Here, the lattice coordination number z is assumed to be equal to 10. Parameters r_i and q_i are, respectively, relative to molecular van der Waals volumes and molecular surface areas of the pure component i. The binary parameters for sugars were taken from the literature [12]. The required van der Waals parameters r_i and q_i of the UNIQUAC model for the ionic liquids were estimated with the correlation proposed by Domanska [23].

$$r_i = 0.029281 \times V_i(cm^3.mol^{-1}), \qquad q_i = \frac{(z-2)\times r_i}{z} + \frac{2}{z} \quad (4)$$

where V_i is the molar volume of the ionic liquid at T = 298.15 K and z is the coordination number assumed to be equal to 10.

To obtain a better description of the (solid + liquid) phase equilibrium simultaneously, temperature-dependent model parameters (Δg_{ij} or Δu_{ij}) were assumed:

$$\Delta g_{12}(J.\ mol^{-1}) = g_{12} - g_{11} = a_{12} + b_{12}\cdot T(\mathrm{K}), \qquad \Delta g_{21}(J.\ mol^{-1}) = g_{21} - g_{22} = a_{21} + b_{21}\cdot T(\mathrm{K}) \quad (5)$$

Each thermodynamic model requires four adjustable parameters per binary (Δg_{ji} or Δu_{ji}). The adjustable parameters were determined by minimization of the following objective function (OF):

$$OF = \sum_{i=1}^{n}(T_{\exp,i} - T_{calc,i})^2 \quad (6)$$

The root-mean-square deviation (RMSD) of temperature, σ_T, was calculated according to the following definition:

$$\sigma_T = \left(\sum_{i=1}^{n}\frac{(T_{\exp,i} - T_{calc,i})^2}{n-2}\right)^{1/2} \quad (7)$$

As can be seen in the results depicted in **Figure 1**, both thermodynamic models correlate well with the experimental data.

http://dx.doi.org/10.5772/67026

2.3. Solubility of carbohydrates in a binary mixture of (IL + EtOH)

2.3.1. Effect of the structure of the carbohydrate

Figure 2 shows the solubility of glucose, fructose, sucrose and lactose in a binary mixture of (BMIMCl + EtOH) for the following experimental conditions: mixture ratio by weight = 10, time = 300 min. The solubility increases in the following order lactose < sucrose < glucose < fructose. This behavior agrees well with the literature [24]. As expected, the solubility of disaccharides in ILs is higher than monosaccharides [15].

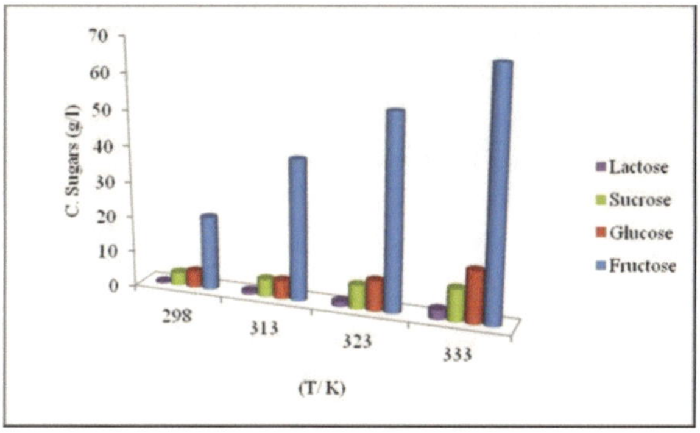

Figure 2. Solubility of carbohydrates in a binary mixture of (BMIMCl + EtOH).

2.3.2. Effect of ethanol/ionic liquid ratio and temperature

Figure 3A presents the effect of (ethanol/ionic liquid) ratio by weight on the solubility of the studied sugars in the binary mixture. It is shown that the solubility of the sugars increases with decreasing (ethanol/ionic liquid) ratio for the studied ionic liquids. This behavior can be related to the polarity and the hydrogen bond basicity of binary mixtures (cosolvent + IL) [25]. Introducing ethanol in IL leads to a reduction in polarity and hydrogen bond. Therefore, the affinity of polar compounds with (cosolvent + BMIMCl) mixtures is lower than in pure IL. A high ethanol/ionic liquid ratio is recommended for the satisfactory extraction of carbohydrates from their mixtures.

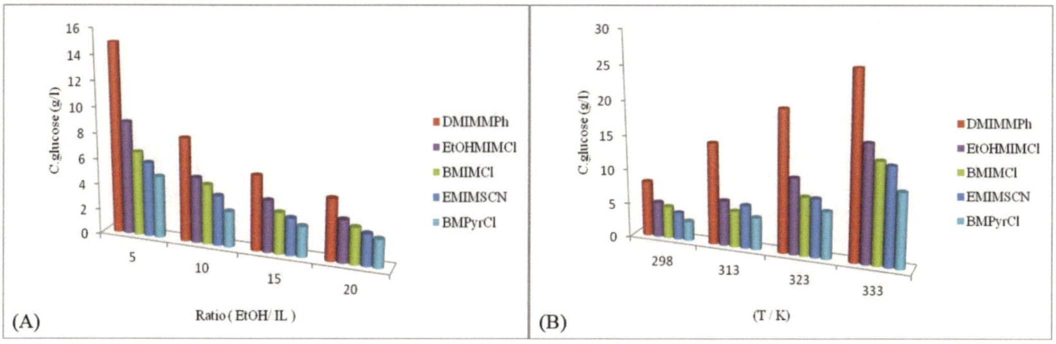

Figure 3. The effect of (A) ethanol/ionic liquid ratio and (B) temperature (K) on the solubility of glucose in the (IL + EtOH) mixture.

On the other hand, the effect of temperature on the solubility of the studied sugars in a mixture of {IL + ethanol} is illustrated in **Figure 3B**. The solubility of carbohydrates is directly proportional to the temperature for the used ionic liquids mixtures. Increasing temperature increases the molecular thermal motion of the entire system and the volatility of solute [26]. Hence, low temperature is highly recommended for a good extraction of carbohydrates.

2.3.3. Applying 2^3 full-factorial design

An experimental design technique, 2^3 full-factorial design, was used to determine the optimal parameters for the extraction of glucose from IL using ethanol and to evaluate the interaction between different parameters. Analysis of variance (ANOVA) data for the system indicates the well fitting of the experimental results to the factorial model equation and hence accuracy of this model. The interaction between the different factors and their effect on the solubility of glucose in this system are shown in **Figure 4**. It is found that there is no interaction between the different factors. As expected, the solubility of glucose increases with increasing temperature and water content and with decreasing the EtOH/IL ratio. The main effective parameter on the glucose solubility is the temperature.

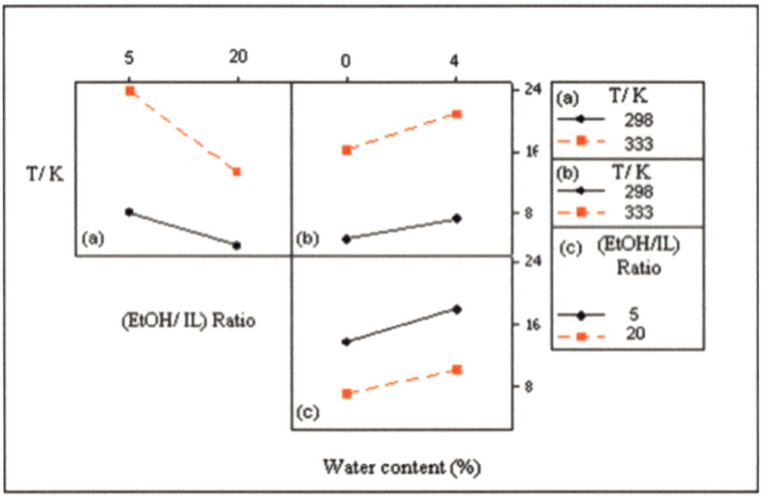

Figure 4. The interaction between different variables on the solubility of glucose in a binary mixture of (BMIMCl + EtOH) at time 300 min: (a) at water content = 2%, (b) at EtOH/IL ratio by weight = 12.5, (c) at temperature = 315.5 K.

2.4. Extraction process using the antisolvent method

Table 2 presents results of the extraction of sugars from five ionic liquids using the antisolvent method with the following characteristics: an {ethanol/ionic liquid} ratio equal to 10, neglected water content and a decrease in temperature from 313 K (or 323 or 333) to 298 K. Results indicate that the extracted percentage increases in the following order fructose < sucrose < glucose < lactose. The performance of different antisolvent on the extraction of sugars from IL using the antisolvent method was evaluated. Results are given in **Table 3**. For these extractions, a solution containing 0.20 wt% of sugar was prepared in BMIMCl at 373 K. After the complete dissolution of sugar, an antisolvent is added at 298 K.

http://dx.doi.org/10.5772/67026

Binary mixture of ethanol + IL	EMIMSCN	DMIMMPh	BMIMCl	EtOHMIMCl	BMPyrCl
Glucose					
Ex.% (from 313 to 298 K)	37.15	45.13	21.91	21.76	39.05
Ex.% (from 323 to 298 K)	53.33	60.39	51.89	53.21	58.63
Ex.% (from 333 to 298 K)	72.08	69.58	72.03	69.39	73.71
Sucrose					
Ex.% (from 313 to 298 K)	29.715	31.593	37.313	24.919	31.283
Ex.% (from 323 to 298 K)	43.223	49.062	52.689	41.640	42.939
Ex.% (from 333 to 298 K)	67.190	65.168	71.286	60.036	66.124

Table 2. The extraction% of the sugars from binary mixtures of the ILs.

Antisolvents	Glucose extracted %	Fructose extracted %	Sucrose extracted %	Lactose extracted %
	BMIMCl			
Ethanol	99	80	99	99
Dichloromethane	99	85	99	99
Acetonitrile	99	80	99	99

Table 3. The extraction% of sugars from BMIMCl with different antisolvents.

To conclude this part, a successful extraction process requires high ethanol/IL ratio, low temperature and low water content. The addition of ethanol in mixtures strongly decreases the solubility of carbohydrates in the binary mixtures. Therefore, ethanol is strongly recommended as an antisolvent for separating sugars from ionic liquids mixtures.

3. Study of the interaction between carbohydrates and ionic liquids using ab initio calculations

Ab initio calculations are used to investigate the fundamental natures of the interaction between carbohydrates and ILs. Results were compared with experimental data.

The ab initio calculations were carried out using GAUSSIAN 98 [27]. The minimum energy geometry was determined by performing calculations with density functional theory (DFT). The hybrid Becke 3-Lee-Yang-Parr, B3LYP, exchange-correlation function with the 6-311+G(d) basis set is employed for the geometry optimizations in this work [28]. Partial atomic charges were derived from the ion pair geometries using the CHELPG method [29].

3.1. Ionic liquids structure optimization and hydrogen bond formation

3.1.1. Optimized structures of ionic liquids

The interaction energy ΔE can be calculated using the following equation [30, 31]:

$$\Delta E(\mathrm{kJmol^{-1}}) = 2625.5 \times [E[\mathrm{cation}]^{+}[\mathrm{anion}]^{-} - (E[\mathrm{cation}]^{+} + E[\mathrm{anion}]^{-})] \quad (8)$$

where $E[\mathrm{cation}]^{+}[\mathrm{anion}]^{-}$ is defined as the total energy of the system and $E[\mathrm{cation}]^{+} + E[\mathrm{anion}]^{-}$ is defined as the sum of the energy of the pure compositions.

The interaction energies are corrected using the basis set superposition error (BSSE) and zero-point energy (ZPE). So, the corrected interaction energy ΔE_{Corr} is calculated as follows [30, 31]:

$$\Delta E_{Corr} = \Delta E + \Delta E_{BSSE} + \Delta E_{ZPE} \quad (9)$$

where ΔE_{BSSE} is the correction of BSSE and ΔE_{ZPE} is the correction of ZPE.

To obtain the stable configurations of DMIMMPh, the anion MPh^{-} is located at several different positions around the cation. The initial configurations are fully optimized. Three representative configurations, A_1, A_2 and A_3, are chosen, where the methyl phosphonate anion (MPh^{-}) is placed at carbon positioned 2, 4 and 5 at the imidazolium ring in dimethyl imidazolium cation ($DMIM^{+}$), respectively. It was found that the interaction energy of A_1 conformer is higher than that of A_2 and A_3 conformers. Therefore, A_1 conformer has better probability than others and is selected as the best optimized structure for DMIMMPh.

The same observation was found with the other ILs, where the best optimized structure was found when the anion was placed at the imidazolium C2. For BMIMCl, the best optimized structure is matched with the work of Hunt et al. [32] who also found that the best optimized structure for BMIMCl is when Cl^{-} is placed at C2 of $BMIM^{+}$. The best optimized structures are shown in **Figure 5**.

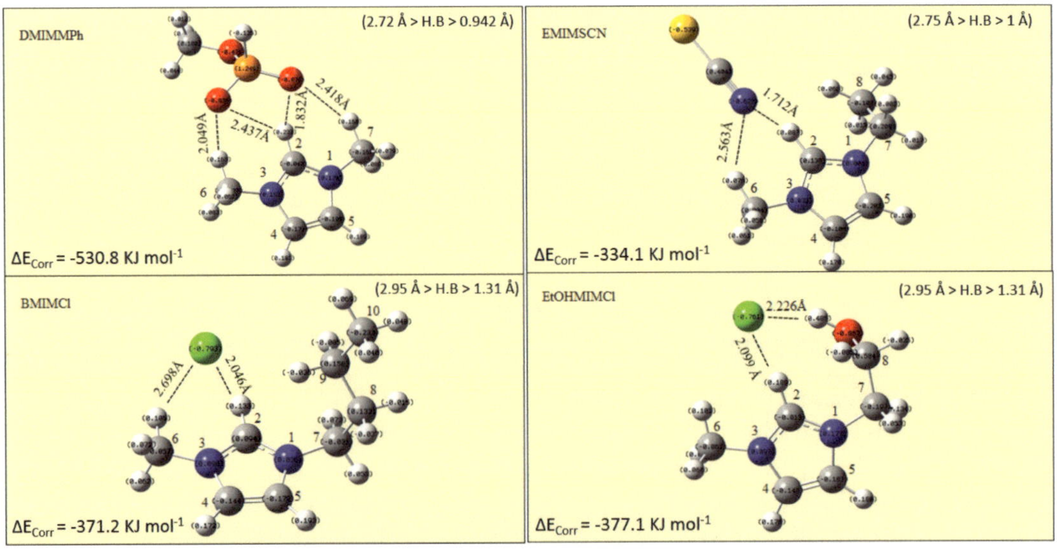

Figure 5. The most stable geometries of DMIMPh, EMIMSCN, BMIMCl, EtOHMIMCl ILs.

In fact, the most stable IL ion pairs have the position of the anion in front of the C2-H. However, the anion is shifted slightly toward the methyl group, the less bulky group, at C6 and lies slightly above the plane of the imidazolium ring.

This behavior is changed in EtOHMIMCl ionic liquid due to the presence of the hydroxyl group at C8. This makes C8 a better hydrogen bond donor than C6. Therefore, the Cl^- anion is shifted slightly toward the OH-C8 and this is matched with the work of Zhang et al. [33].

Ab initio calculation results for the interaction energies of the optimized ILs showed that ΔE_{Corr} increases in the following order: EMIMSCN < BMIMCl ≈ EtOHMIMCl < DMIMMPh (**Figure 5**).

3.1.2. Hydrogen-bonding interaction

It is now well established that cation-anion H bonding occurs in some imidazolium-based ionic liquids [34]. Jeffrey found that a normal O···H hydrogen bond is formed when O···H < 2.7 Å and C– H···O > 90° [35]. Otherwise, a criterion that the interaction distance be less than the sum of the respective van der Waals radii has also been used.

In case of BMIMCl, the Cl···H bond length is between 2.05 and 2.67 Å. This distance is longer than the covalent bond length in H···Cl (1.31 Å), but it is shorter than the van der Waals distance of Cl···H (2.95 Å). These values approve that the Cl^- anion could form a hydrogen bond with the N–C–H fragments of $BMIM^+$ cation (2.95 Å > H.B > 1.31 Å).

Ab initio calculations show the formation of two hydrogen bonds between $BMIM^+$ cation and Cl^- anion. The obtained results are matched with those found in the literature [32]. Hunt et al. [32] proved that a strong hydrogen bond could be formed when the C–H···Cl length is lower than 2.3 Å, while a weak hydrogen bond is formed if $2.3 < r < 2.75$ Å.

Also, the same behavior was observed with the O—H bond length in DMIMMPh ionic liquid where: 2.72 Å > H.B > 0.942 Å and also the N—H bond length in EMIMSCN ionic liquid where: 2.75 Å > H.B > 1 Å.

On the other hand, a strong hydrogen-bond interaction is observed between the Cl^- anion and the hydroxyl group on the cation (O–H—Cl) (**Figure 5**).

3.2. Interaction of ionic liquids with carbohydrates

Results concerning the binary systems {glucose + IL} show that the interaction of glucose-IL systems increases in the following order: glucose-BMIMCl < glucose-DMIMMPh (**Figure 6**). This observation is in good agreement with the experimental results obtained.

On the other hand, the cellulose-building unit, cellobiose, interacts similarly to glucose with anions and cations of ionic liquids (**Figure 7**). The bond strength (Å) cellulose-IL systems show that cellulose interacts with ionic liquids mainly via hydrogen-bonding formation. The interaction energies and the hydrogen-bonding formation of cellulose-IL systems increase in the following order: cellulose-BMIMCl < cellulose-DMIMMPh. This observation is in good agreement with the experimental results obtained in this work.

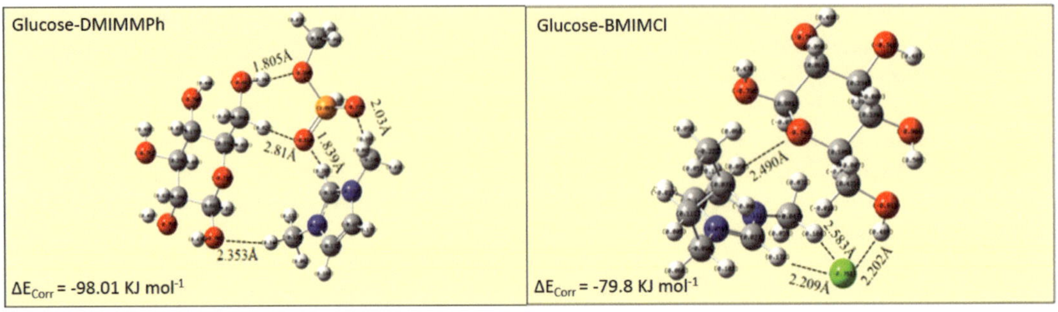

Figure 6. The interaction of glucose with (A) DMIMMPh and (B) BMIMCl.

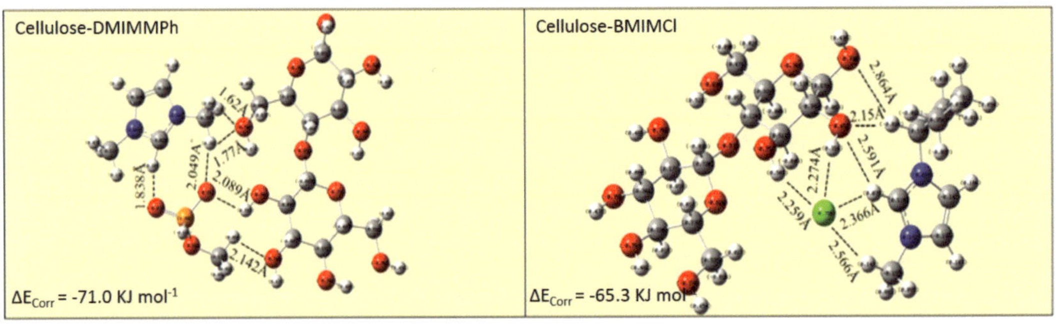

Figure 7. The interaction of cellulose with (A) DMIMMPh and (B) BMIMCl.

Comparing the values of both the interaction energy and the hydrogen bond strength of {glucose-Ils} and {cellulose-Ils} systems proved that ILs interact with glucose higher than with cellulose. This could be a result of the strong β-(1→4)-glycosidic linkages between cellulose molecules compared to the linkages between glucose molecules.

3.3. Experimental study on the dissolution and regeneration of cellulose

Solid-liquid equilibria (SLE) of microcrystalline cellulose (MCC) in ionic liquids were carried out in a large range of temperature (313–393 K). Ab initio calculation results proved that ionic liquids are capable of dissolving cellulose molecules by breaking down the inter- and intramolecular hydrogen bond's network between cellulose molecules and setting up a new IL-cellulose hydrogen bond's network promoting their dissolution.

The dissolution of cellulose is higher in DMIMMPh ionic liquid, up to 13 mass fraction at 393 K, than other ionic liquids and this is matched with the theoretical results. This is maybe related to its high interaction energy, strong hydrogen bond formation, with its high hydrogen bond basicity and polarity.

It could be concluded from the analysis results, **Figures 8–10**, that DMIMMPh ionic liquid is an excellent solvent for the regeneration of cellulose from biomass feedstocks. The recovered cellulose is more accessible to the enzymes, for example, cellulose enzyme, which convert cellulose and hemicelluloses into fermentable sugars for biofuel production. These results also proved that there are no chemical reactions occurred during the dissolution and regeneration process.

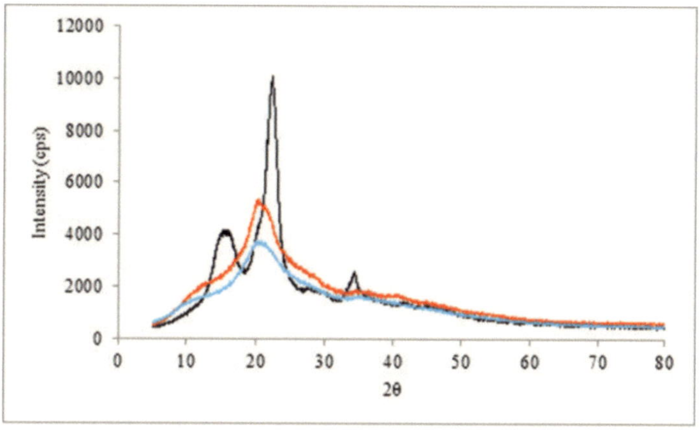

Figure 8. XRD of (■) MCC and cellulose regenerated in (■) BMIMCl and (■) DMIMMPh.

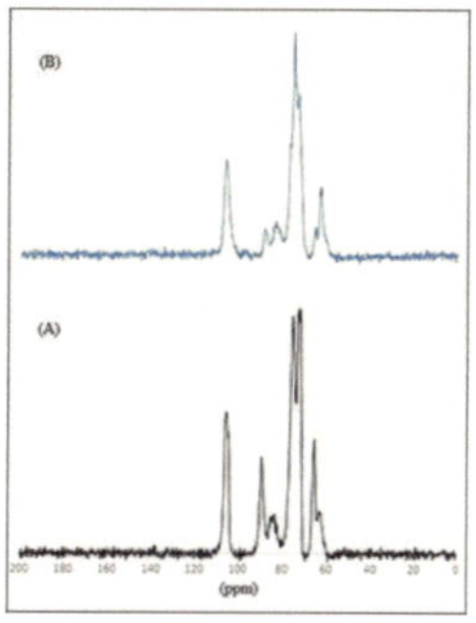

Figure 9. ^{13}C CP/MAS NMR spectra of (A) MCC and (B) regenerated cellulose.

Figure 10. SEM images of (A) MCC and (B) regenerated cellulose.

4. Pretreatment of miscanthus using 1,3-dimethyl-imidazolium methyl phosphonate (DMIMMPh) ionic liquid for glucose recovery and ethanol production

Miscanthus is a promising biomass energy crop due to its relatively low maintenance and high energy content [6]. This part presents an environmentally friendly method of extracting cellulose from miscanthus using DMIMMPh ionic liquid and also provides a new approach for utilizing biomass resources.

4.1. Miscanthus solubility in ionic liquids

Imidazolium-based ionic liquids have the ability to dissolve complex macromolecules such as cellulose with high dissolution efficiency by breaking their inter- and intramolecular hydrogen bond's network [9]. The solid-liquid equilibria of {miscanthus + ILs} system were carried out in a large temperature's range (90–130°C). It is investigated that DMIMMPh has the higher dissolution efficiency than other ionic liquids **Figure 11**. This could be related to its high hydrogen bond basicity and polarity compared to other ILs. This indicates that the nature and the size of the ionic liquid's anion greatly influence the solubility of miscanthus. Strong H-bond-acceptor anions, such as phosphonate anions, effectively dissolve miscanthus. The high solubility of miscanthus in phosphonate-based ILs (DMIMMPh) may be of interest in technology of pretreatment of biomass because these ILs are free of halogens. Moreover, these ILs are less toxic, noncorrosive and biodegradable [9, 14].

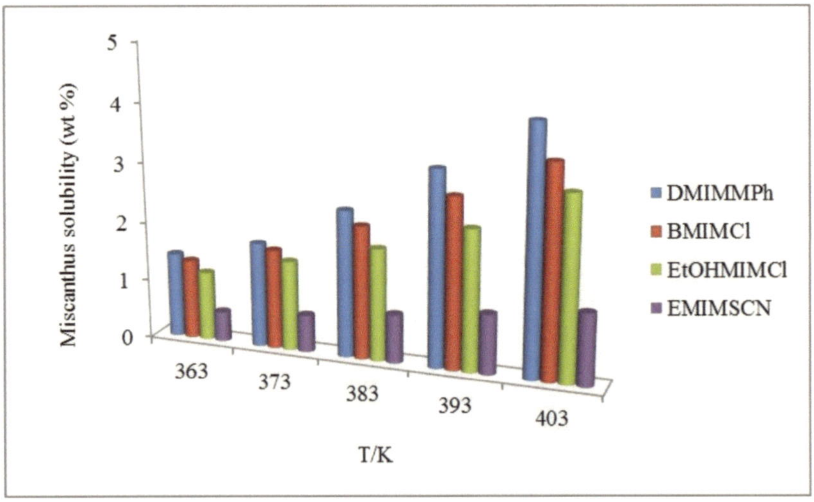

Figure 11. The solubility of miscanthus in ionic liquids.

It is suggested that, in salt solutions with small, strong polarizing cations and large polarizable anions, intensive interactions with cellulose, occur. Results obtained on the solubility of miscanthus in ionic liquids and the data published in the literature [8, 9, 14, 36] proved

http://dx.doi.org/10.5772/67026

that phosphonate- and chloride-based ionic liquids are good solvents for miscanthus. The presented solubility results are matched with those of Padmanabhan et al. [14].

4.2. Cellulose extraction using DMIMMPh ionic liquid

4.2.1. Effect of temperature and time

Brandt et al. [37] found that swelling and dissolution in ILs are temperature dependent and that better dissolution and regeneration rates are obtained at temperatures beyond 373 K.

Figure 12A presents the dissolution and regeneration rates as well as the cellulose grade and recovery for the {miscanthus + DMIMMPh} system with 5% miscanthus mass fraction during 6 h and using water as an antisolvent.

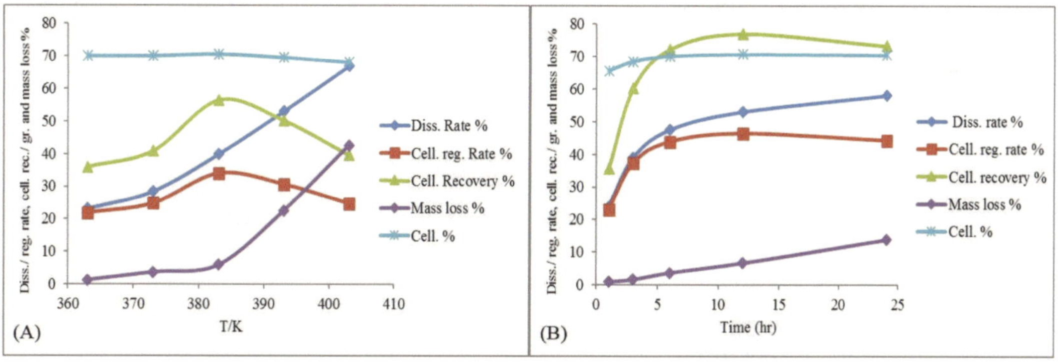

Figure 12. The effect of (A) temperature (K) and (B) time (h) on the extraction process.

Increasing the temperature, from 363 to 403 K, increases the dissolution rate and cellulose grade. Also, increasing the temperature, from 363 to 383 K, increases the cellulose recovery and regeneration rate. But at temperature higher than 383 K, cellulose recovery and regeneration rate decrease obviously due to the degradation occurrence. Mass loss proved that temperature has a significant influence on the extraction efficiency. The mass loss increased up to 30 times when increasing reaction time from 1 to 24 h. High temperature ranges, more than 380 K, lead to high degradation rates and low cellulose extraction efficiency [36].

The kinetics of the miscanthus dissolved in DMIMMPh were studied as a function of dissolution time. **Figure 12B** shows the dissolution rate%, cellulose recovery%, cellulose grade%, cellulose regeneration rate% and mass loss% miscanthus in DMIMMPh. The measurements were performed in these conditions: at 3% miscanthus mass fraction, temperature of 373 K and water is used as an antisolvent. It is obvious that increasing reaction time from 1 to 24 h increases the dissolution rate. Also, increasing reaction time from 1 to 12 h significantly increases cellulose grade, recovery and regeneration rate. On the other hand, increasing reaction time more than 12 h, the cellulose grade, recovery and regeneration rate, decreases due to the degradation occurrence. The mass loss increased up to 14 times when increasing reaction time from 1 to 24 h. Hence, polymer degradation increases with time.

4.2.2. Applying Box-Behnken experimental design on the extraction process

The results of miscanthus solubility and cellulose regeneration in DMIMMPh were evaluated using Box-Behnken experimental design. This experimental design allows the study of the effect of each factor temperature, time and miscanthus mass fraction, as well as the effects of interactions between factors on the cellulose solubility and regeneration.

According to this design, the optimal conditions were estimated using a second-order polynomial function by which a correlation between studied factors and response (mean diameter) was generated. The general form of this equation is as follows:

$$Y = \beta_o + \beta_1 X_1 + \beta_2 X_2 + \beta_3 X_3 + \beta_{12} X_1 X_2 + \beta_{13} X_1 X_3 + \beta_{23} X_2 X_3 + \beta_{11} {X_1}^2 + \beta_{22} {X_2}^2 + \beta_{33} {X_3}^2 \quad (10)$$

where Y is the predicted response; dissolution rate, cellulose grade and cellulose recovery%, X_1, X_2 and X_3, are studied variables; temperature, time and miscanthus mass fraction; β_{ij} are equation constants and coefficients.

The analysis of variance data, ANOVA, for the system indicates the well fitting of the experimental results to the polynomial model equation and hence accuracy of this model.

Figure 13 presents the plots of the response surface of cellulose recovery. It is shown that increasing both temperature and time increase the dissolution rate and cellulose recovery. High miscanthus mass fraction decreases the cellulose recovery. No important interactions between parameters were observed. Temperature is the most effective parameter on the cellulose dissolution and regeneration.

The optimum conditions obtained from the Box-Behnken design for {miscanthus + DMIMMPh} mixture are 3.2% miscanthus mass fraction and heating up to 107°C during 6.30 h. Applying these optimum conditions resulted in the regeneration of cellulose with grade of 74.4% and recovery of 77.6% (**Table 4**). A slight decrease in efficiency was observed when using recycled DMIMMPh (**Table 4**).

4.3. Characterization of the regenerated cellulose

Analyses results, **Figures 14–17**, indicate that the regenerated cellulose-rich extract is homogeneous, dense and porous and has a higher surface area. Analyses investigations evidenced the production of amorphous cellulose almost free of lignin, which is suitable for enzymatic hydrolysis processes.

4.4. Production of bioethanol

4.4.1. Enzymatic hydrolysis

The regenerated cellulose was hydrolyzed enzymatically to produce glucose. The optimum conditions used for the enzymatic hydrolysis process were heating up to 50°C during 72 h and adding cellulase enzyme at loading 20 FPU/g of cellulose with 1:1 volume of β-glucosidase.

http://dx.doi.org/10.5772/67026

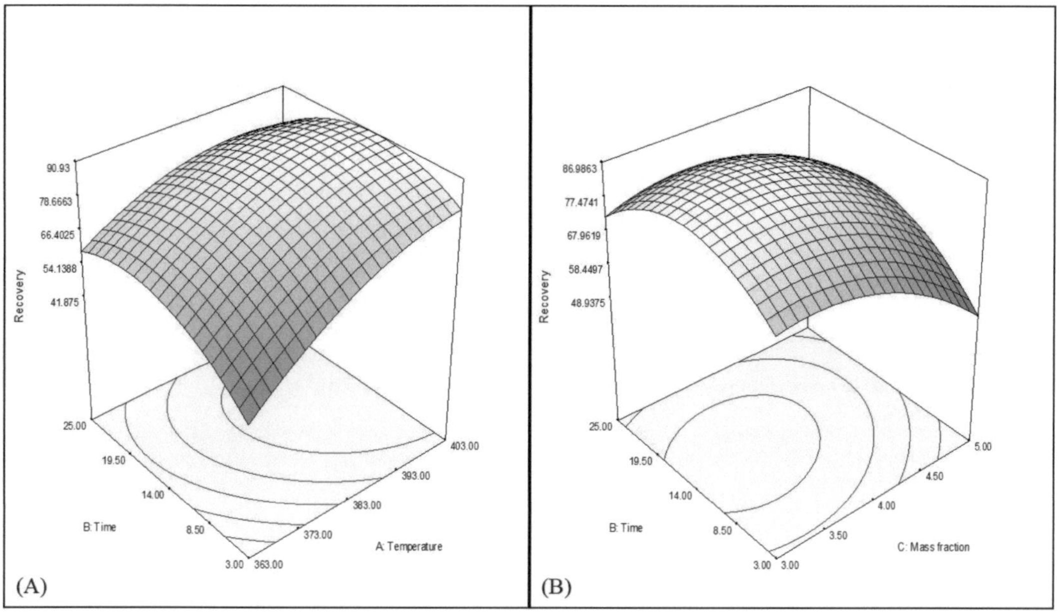

Figure 13. The response surface plots of cellulose recovery for miscanthus-DMIMMPh mixture (A) at mass fraction 4% and (B) at temperature 110°C.

Solvent	Cellulose grade %	Cellulose recovery %	Hemicell. grade %	Hemicell. recovery %	Lignin grade %	Lignin recovery %	Others (ash and unknown products) %
DMIMMPh	74.4	77.6	13.7	22.9	9.1	14.4	2.80
BMIMCl	71.8	73.8	14.9	24.6	10.5	16.4	2.90
Recycled DMIMMPh	72.3	73.5	15.1	24.9	9.7	15.0	3.00

Table 4. Applying the optimum parameters of Box-Behnken design for the miscanthus-DMIMMPh and BMIMCl mixtures.

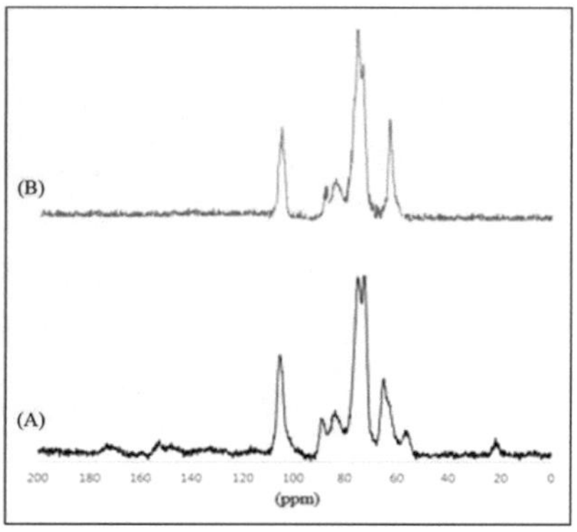

Figure 14. NMR spectra of miscanthus: (A) untreated and (B) treated with DMIMMPh IL.

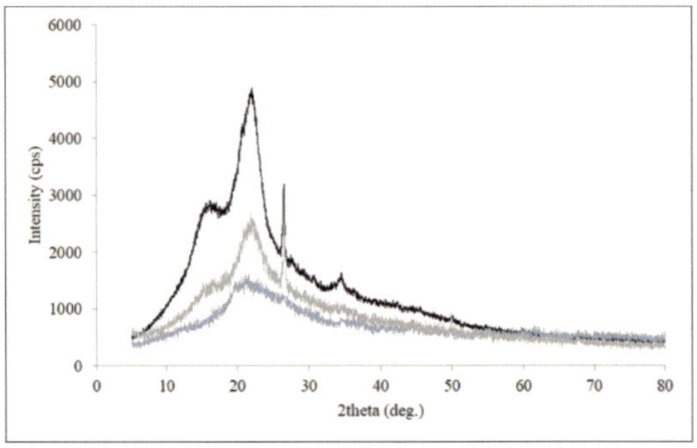

Figure 15. XRD of (■) RD miscanthus, (■) miscanthus residue and (■) cellulose-rich extract.

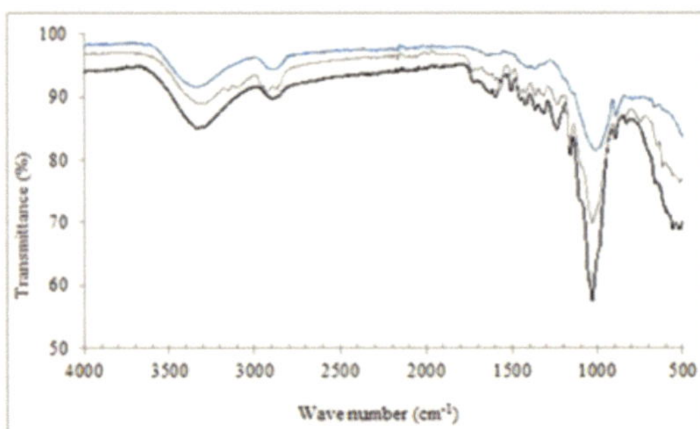

Figure 16. IR of (■) R miscanthus, (■) miscanthus residue and (■) cellulose-rich extract.

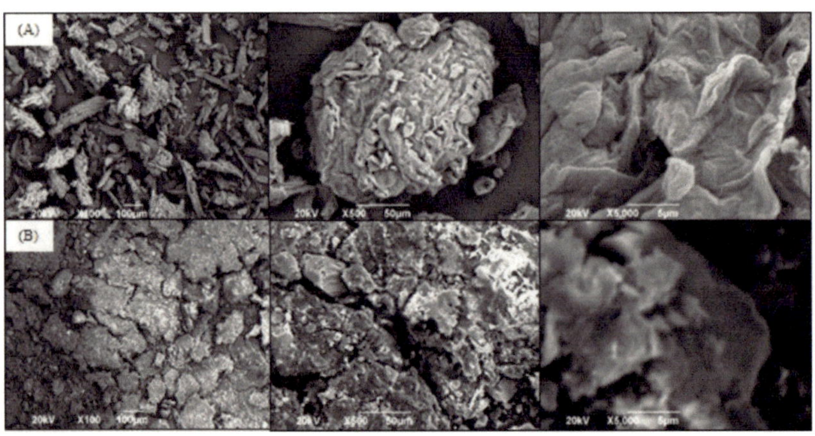

Figure 17. SEM images of (A) MCC cellulose and (B) the cellulose-rich extract with DMIMMPh.

Figure 18 represents the hydrolysis rate for the original miscanthus and the regenerated cellulose. It is obvious that the enzymatic hydrolysis rate decreases rapidly. The glucose hydrolysis efficiency reaches up to 94%.

Table 5 shows the results of different hydrolysis processes for untreated miscanthus and regenerated cellulose. Glucose hydrolysis efficiency of the untreated sample is only 11.9%, while it is more than 90% for the regenerated cellulose samples. The hydrolysis efficiency of the samples treated with DMIMMPh is higher than those treated with BMIMCl. This is maybe due to the inhibition effect of the chloride ions [14] on the cellulase enzyme.

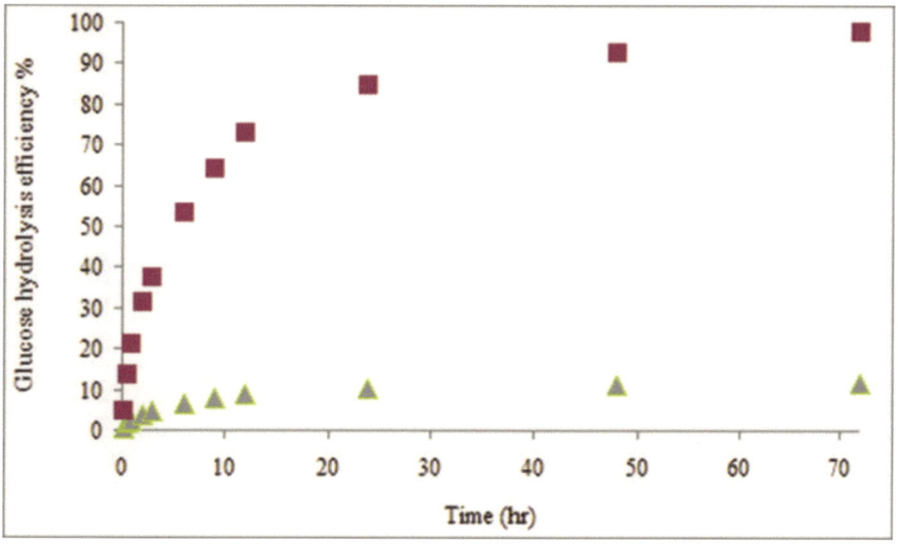

Figure 18. Glucose hydrolysis efficiency as a function of time for: (▲) untreated miscanthus, (■) cellulose regenerated with DMIMMPh.

Sample	Glucose (g/l)	Glucose hyd. effic. (%)	Xylose (g/l)	Xylose hyd. effic. (%)	Ethanol (g/l)	Ethanol conversion rate (%)	Ethanol production effic. (%)	Ethanol produced in g/kg miscanthus
Original miscanthus	1.39	11.89	0.5	6.80	0.53	55.02	4.51	17.70
DMIMMPh	41.88	93.98	5.02	61.18	20.40	85.13	66.75	148.43
Recycled DMIMMPh	37.92	87.77	3.35	43.58	16.72	79.28	55.88	119.14
BMIMCl	32.53	84.72	3.1	45.52	14.72	80.85	49.03	115.63

Table 5. Results for different hydrolysis and fermentation processes for untreated miscanthus and regenerated cellulose samples.

Furthermore, the regenerated cellulose samples have a low content of hemicellulose which could be converted into xylose during the hydrolysis process. The xylose conversion efficiency does not exceed 70%.

4.4.2. Hydrolysate fermentation

The hydrolysate solutions of the untreated miscanthus and the regenerated cellulose samples were fermented using yeast from *S. cerevisiae*. **Figure 19** shows the results of fermentation of the hydrolysate produced from the treatment of miscanthus with DMIMMPh. These results include the sugars consumption, cell growth and ethanol formation.

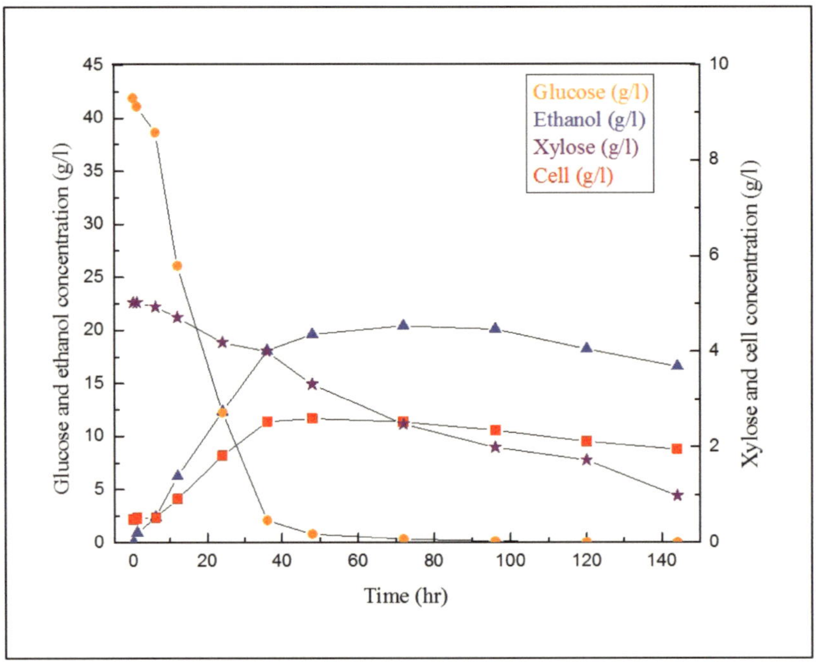

Figure 19. Fermentation results and bioethanol production.

The results indicate the formation of a typical batch growth phase. This batch includes four phases as following: lag phase (0–6 h), exponential growth phase (6–36 h), deceleration phase (36–48 h) and stationary phase (48–72 h).

At the beginning of the fermentation process, lag phase, the yeasts adapted themselves to the growth conditions and hence, the concentration of glucose is quite constant during the first 6 h. Then, after 6 h, exponential phase, both the substrate consumption and cell growth are increasing greatly with time. During the declaration phase, 36–48 h, it is observed that the growth rate is slowing down due to the glucose depletion. As expected, decreasing the glucose concentration during the fermentation process is followed by an increase in cell and ethanol production. Hence, ethanol concentration increases greatly during the exponential and declaration and stationary phases and achieves a maximum concentration of 20.35 g/L after 72 h. Then, the ethanol concentration is decreased as it is used as a carbon source for the yeast growth when the glucose concentration is depleted [38]. The cell growth rate (u) is 0.0677 per hour. This result is matched with the work of Cheng et al. [39].

Table 5 shows the results of the fermentation conversion rate of different hydrolysates with a maximum value of 85%. It also shows the ethanol production efficiency for carbohydrate content, cellulose and hemicellulose, in the regenerated cellulose which is calculated as following [40–42]:

$$\text{Ethanol production efficiency}(\%) = 100\times\left\{\frac{\textit{ethanol produced}(g)}{0.511\times\{1.11\times \textit{cellulose content}(g)+1.136\times \textit{hemicell.content}(g)\}}\right\} \quad (11)$$

where the value 1.11 is equivalent to [molecular weight of glucose unit (180.16)/molecular weight of cellulose unit (162.14)], while the value 1.136 is equivalent to [molecular weight of xylose unit (150.13)/molecular weight of hemicellulose unit, xylan, (132.11)].

The obtained ethanol production efficiency is 66.75% **Table 5**. These values could be compared with those published in the literature for the hydrolysates based on acid and alkali pretreatments in which the ethanol production efficiency ranges from 65 to 75% [40–42].

Finally, an overall ethanol yield was calculated in order to have a complete evaluation for the production of bioethanol from miscanthus. This calculation is based on total amount of miscanthus considering the loss of carbohydrates during the pretreatment, enzymatic hydrolysis and fermentation processes. The overall ethanol yield for the miscanthus treated with DMIMMPh was 148 g ethanol per kg miscanthus. This value indicates the high efficiency of the presented process when compared the values published in the literature for other biomass feedstocks [40–42].

To conclude this part, the use of DMIMMPh ionic liquid in the pretreatment of miscanthus and the extraction of cellulose has been studied. Analyses results evidenced the production of amorphous, porous cellulose almost free of lignin. The glucose hydrolysis efficiency of the regenerated cellulose reached up to 94%. The IL pretreatment efficiency is determined by its ability to improve cellulose accessibility and increase overall sugars. This is confirmed with the high overall ethanol yield, up to 148 g ethanol kg^{-1} miscanthus, produced from the fermentation of the hydrolysates.

5. Conclusions

The solubility of glucose, fructose, sucrose and lactose in ionic liquids was measured within a temperature range from 283 to 383 K. It is observed that the solubility of disaccharides exhibits a noticeably lower solubility than monosaccharides. Solubility data were successfully correlated with NRTL and UNIQUAC thermodynamic models. The possibility of extracting sugars from ILs using the antisolvent method has been evaluated. A successful extraction process requires high ethanol/IL ratio, low temperature and low water content. The considerable decrease in the solubility of sugars in the binary mixtures proves the ability of ethanol as an excellent antisolvent [43, 44].

Ab initio calculations were used as a tool to investigate the fundamental natures of the interaction between carbohydrates and ILs. The most stable geometries ILs have been studied. The most stable IL ion pairs have the anion positioned in front of the C2-H of the imidazolium ring of cation. It was concluded that the anion plays the main role in the dissolution process of carbohydrates, in which the H-bonding forces are the major interactions. The obtained results proved that DMIMMPh is more efficient than other ionic liquids. The analyses results indicate that the regenerated cellulose is quite similar to the original one with a great decrease in its crystallinity. This proves that the treatment of cellulose with ILs is accompanied only with a physical change [45].

The use of ILs in the pretreatment of miscanthus has been studied. Extraction results were evaluated using Box-Behnken design. Analyses results evidenced the production of amorphous, porous cellulose almost free of lignin, thereby facilitating its enzymatic hydrolysis. The glucose hydrolysis efficiency of the regenerated cellulose reached up to 97.74%. The pretreatment efficiency is determined by its ability to improve cellulose accessibility and increase overall sugars yield. This could be confirmed with the high overall ethanol yield, up to 150 g ethanol kg^{-1} miscanthus, produced from the fermentation of the hydrolysates. Alkylphosphonate anion-based ILs were proposed as excellent candidates for biomass treatment and biofuel production [46, 47].

Author details

El-Sayed R.E. Hassan[1,2] and Fabrice Mutelet[1]*

*Address all correspondence to: fabrice.mutelet@univ-lorraine.fr

1 Reactions and Process Engineering Laboratory (CNRS UMR 7274), University of Lorraine, Nancy, France

2 Central Metallurgical Research and Development Institute (CMRDI), Helwan, Cairo, Egypt

References

[1] Tadesse H., Luque R., Advances on biomass pretreatment using ionic liquids: an overview. Energy Environ Sci., (2011); **4**:3913–3929.

[2] Henry R., Evaluation of plant biomass resources available for replacement of fossil oil. Plant Biotechnol J., (2010); **8**:288–293.

[3] Balat M., Balat H., Oz C., Progress in bioethanol processing. Prog Energy Combust Sci., (2008); **34**:551–573.

[4] Agbor V. B., Cicek N., Sparling R., Berlin A., Levin D. B., Biomass pretreatment: fundamentals toward application. Biotechnol Adv., (2011); **29**:675–685.

[5] Wang H., Gurau G., Rogers R. D., Ionic liquid processing of cellulose. Chem Soc Rev., (2012); **41**:1519–1537.

[6] Brosse N., Dufour A., Meng X., Sun Q., Ragauskas A., Miscanthus: a fast growing crop for biofuels and chemicals production. Biofuels Bioprod Bioref., (2012); **6**:580–598.

[7] Dadi A. P., Schall C. A., Varanasi S., Mitigation of cellulose recalcitrance to enzymatic hydrolysis by IL pretreatment. Appl Biochem Biotechnol., (2007); **136–140**: 407–421.

[8] Sun N., Rahman M., Qin Y., Maxim M. L., Rodríguez H., Rogers R., Complete dissolution and partial delignification of wood in the ionic liquid 1-ethyl-3-methylimidazolium acetate. Green Chem., (2009); **11**:646–655.

[9] Swatloski R. P., Spear S. K., Holbrey J. D., Rogers R. D., Dissolution of cellulose with ionic liquids. J Am Chem Soc., (2002); **124**:4974–4975.

[10] Liu Q. B., Janssen M. H. A., van Rantwijk F., Sheldon R. A., Room-temperature ionic liquids that dissolve carbohydrates in high concentrations. Green Chem., (2005); **7**:39–42.

[11] Carneiro A. P., Rodriguez O., Macedo E. A., Solubility of monosaccharides in ionic liquids —experimental data and modeling. Fluid Phase Equilib., (2012); **314**:22–28.

[12] Xu H., Pan W., Wang R., Zhang D., Liu C., Understanding the mechanism of cellulose dissolution in 1-butyl-3-methyl imidazolium chloride ionic liquid via quantum chemistry calculations and molecular dynamics simulations. J Comput Aid Mol Des, (2012); **26**: 329–337.

[13] Shill K., Padmanabhan S., Xin Q., Prausnitz J. M., Clark D. S., Blanch H. W., Ionic liquid pretreatment of cellulosic biomass: enzymatic hydrolysis and ionic liquid recycle. Biotechnol Bioeng., (2011); **108**:511–520.

[14] Padmanabhan S., Kim M., Blanch H. W., Prausnitz J. M., Solubility and rate of dissolution for Miscanthus in hydrophilic ILs. Fluid Phase Equilib., (2011); **309**:89–96.

[15] Conceiçao L. J. A., Bogel-Lukasik E., Bogel-Lukasik R., A new outlook on solubility of carbohydrates and sugar alcohols in ionic liquids. RSC Adv., (2012); **2**:1846–1855.

[16] Fort D. A., Swatolski R. P., Moyna P., Rogers R. D., Moyna G., Use of ionic liquids in the study of fruit ripening by high-resolution 13C NMR spectroscopy: 'Green' solvents meet green bananas. Chem Commun., (2006); **7**:714–716.

[17] Zhang H., Gurau G., Rogers R. D., Ionic liquid processing of cellulose. Chem Soc Rev., (2012); **41**:1519–1537.

[18] Malgorzata E. Z., Ewa B. L., Rafal B. L., Solubility of carbohydrates in ionic liquids. Energy Fuels, (2010); **24**:737–745.

[19] Renon H., Prausnitz J. M., Local compositions in thermodynamic excess functions for liquid mixtures. AIChE J., (1968); **14**:135–144.

[20] Abrams D. S., Prausnitz J. M., Statistical thermodynamics of liquid mixtures: a new expression for the excess gibbs energy of partly or completely miscible systems. AIChE J., (1975); **21**:116–128.

[21] Prausnitz J. M., Lichtenthaler R. N., Azevedo E. G., Molecular thermodynamics of fluid-phase equilibria, 2nd ed.; Prentice-Hall Inc.; Englewood Cliffs, NJ, 1986.

[22] Simoni L. D., Chapeaux A., Brennecke J. F., Stadtherr M. A., Asymmetric framework for predicting liquid-liquid equilibrium of ionic liquid-mixed-solvent systems. 2. Prediction of ternary systems. Ind Eng Chem Res., (2009); **48**:7257–7265.

[23] Domanska U., Solubility of n-alkanols (C16, C18, C20) in binary solvent mixtures. Fluid Phase Equilib., (1989); **4**:6223–248.

[24] Hyvonen L ., Koivistoinen P., Fructose in food systems. Applied Science Publishers. : In Birch, G.G. & Parker, K.J. Nutritive Sweeteners. London & New Jersey; pp. 133–144. ISBN 0-85334-997-5, 1982.

[25] Yang Q., Xing H., Su B., Yu K., Bao Z., Yang Y., et al., Improved separation efficiency using ionic liquid–cosolvent mixtures as the extractant in liquid–liquid extraction: a multiple adjustment and synergistic effect. Chem Eng J., (2012); **181–182**: 334–342.

[26] Liu W., Hou Y., Wu W., Ren S., Jing Y., Zhang B., Solubility of glucose in ionic liquid + antisolvent mixtures. Ind Eng Chem Res., (2011); **50**: 6952–6956.

[27] Frisch M. J., Trucks G. W., Schlegel H. B., Scuseria G. E., Robb M. A., Cheeseman J. R., et al., Gaussian 03, revision E.01, Gaussian, Inc.: Pittsburgh, PA, 2003.

[28] Becke A. D., Density-functional thermochemistry. III. The role of exact exchange. J Chem Phys., (1993); **98**:5648–5652.

[29] Breneman C. M., Wiberg K. B., Determining atom-centered monopoles from molecular electrostatic potentials: the need for high sampling density in formamide conformational analysis. J Comp Chem., (1990); **11**:361–373.

[30] Ji W., Ding Z., Liu J., Song Q., Xia X., Gao H., et al., Mechanism of lignin dissolution and regeneration in ionic liquid. Energy Fuels, (2012); **26**:6393–6403.

[31] Zhou J., Mao J., Zhang S., Ab initio calculations of the interaction between thiophene and ionic liquids. Fuel Process Technol., (2008); **89**:1456–1460.

[32] Hunt P. A., Gould I. R., Structural characterization of the 1-butyl-3-methylimidazolium chloride ion pair using ab initio methods. J Phys Chem A, (2006); **110**:2269–2282.

[33] Zhang S., Qi X., Ma X., Lu L., Zhang Q., Deng Y., Investigation of cation-anion interaction in 1-(2-hydroxyethyl)-3-methylimidazolium-based ion pairs by density functional theory calculations and experiments. J Phys Org Chem., (2012); **25**:248–257.

[34] Chiappe C., Pieraccini D., Ionic liquids: solvent properties and organic reactivity. J Phys Org Chem., (2005); **18**:275–297.

[35] Jeffrey G. A., The ups and downs of C–H hydrogen bonds. J Mol Structure, (1999); **485–486**:293–298.

[36] Wanga X., Li H., Cao Y., Tang Q., Cellulose extraction from wood chip in an ionic liquid 1-allyl-3-methylimidazolium chloride. Bioresour Technol., (2011); **102**:7959–7965.

[37] Brandt A., Hallett J. P., Leak D. J., Murphy R. J., Welton T., The effect of the ionic liquid anion in the pretreatment of pine wood chips. Green Chem., (2010); **12**:672.

[38] Coppella S. J., Dhurjati P., A detailed analysis of *Saccharomyces cerevisiae* growth kinetics in batch, fed-batch and hollow-fiber bioreactors. Chem Eng J., (1989); **41**:B27–B35.

[39] Cheng N. G., Hasan M., Kumoro A. C., Ling C. F., Tham M., Production of ethanol by fed-batch fermentation. Pertanika J Sci Technol., (2009); **17**:399–408.

[40] Yue Z., Teater C., MacLellan J., Liu Y., Liao W., Development of a new bioethanol feedstock-Anaerobically digested fiber from confined dairy operations using different digestion configurations. Biomass Bioenergy, (2011); **35**:1946–1953.

[41] Yue Z., Teater C., Liu Y., MacLellan J., Liao W., A sustainable Pathway of Cellulosic Ethanol production integrating anaerobic digestion with biorefining. Biotechnol Bioeng, (2010); **105**:1031–1039.

[42] Tan L., Tang Y-Q., Nishimura H., Takei S., Morimura S., Kida K., Eficcient production of bioethanol from corn stover by pretreatment with a combination of sulfuric acid and sodium hydroxide. Prep Biochem Biotechnol, (2013); **43**:682–695.

[43] Hassan El-Sayed R. E., Mutelet F., Pontvianne S., Moise J-C., Studies on the dissolution of glucose in ionic liquids and extraction using the antisolvent method. Environ Sci Technol., (2013); **47**: 2809–2816.

[44] Hassan El-Sayed R. E., Mutelet F., Moise J-C., From the dissolution to the extraction of carbohydrates using ionic liquids. RSC Adv., (2013); **3**:20219–20226.

[45] Hassan El-Sayed R. E., Mutelet F., Bouroukba M., Experimental and theoretical study of carbohydrate-IL interactions. Carbohydr Polym, (2015); **127**:316–324.

[46] Hassan El-Sayed R. E., Mutelet F., Moise J-C., The pretreatment of miscanthus using ionic liquids; a way for biofuel production, MATEC Web Conf., (2013); **3**:01049.

[47] Hassan El-Sayed R. E., Mutelet F., Moise J-C., Brosse N., Pretreatment of miscanthus using 1,3-dimethyl-imidazolium methyl phosphonate (DMIMMPh) ionic liquid for glucose recovery and ethanol production. RSC Adv., (2015); **5**:61455–61464.